Quality Gaging Tips

By George Schuetz and Jim McCusker
Mahr Federal, Inc.

Hanser Publishers, Munich

HANSER
Hanser Publications, Cincinnati

Distributed in North and South America by:
Hanser Publications
6915 Valley Avenue, Cincinnati, Ohio 45244-3029, USA
Fax: (513) 527-8801
Phone: (513) 527-8977
www.hanserpublications.com

Distributed in all other countries by
Carl Hanser Verlag
Postfach 86 04 20, 81631 München, Germany
Fax: +49 (89) 98 48 09
www.hanser.de

The use of general descriptive names, trademarks, etc., in this publication, even if the former are not especially identified, is not to be taken as a sign that such names, as understood by the Trade Marks and Merchandise Marks Act, may accordingly be used freely by anyone.
While the advice and information in this book are believed to be true and accurate at the date of going to press, neither the authors nor the editors nor the publisher can accept any legal responsibility for any errors or omissions that may be made. The publisher makes no warranty, express or implied, with respect to the material contained herein.

Library of Congress Cataloging-in-Publication Data

Schuetz, George, 1954-
 Quality gaging tips / by George Schuetz and Jim McCusker.—1st ed.
    p. cm.
  ISBN 1-56990-391-3
 1. Gages. 2. Gaging. I. McCusker, Jim, 1956-1993. II. Title.
 TJ1166.S38 2006
 681'.2—dc22
                        2005033071

Bibliografische Information Der Deutschen Bibliothek
Die Deutsche Bibliothek verzeichnet diese Publikation in der Deutschen Nationalbibliografie; detaillierte bibliografische Daten sind im Internet über <http://dnb.d-nb.de> abrufbar.

ISBN 978-3-446-41976-6

All rights reserved. No part of this book may be reproduced or transmitted in any form or by any means, electronic or mechanical, including photocopying or by any information storage and retrieval system, without permission in writing from the publisher.

© Carl Hanser Verlag, Munich 2011
Reprint of the 1st Edition © 2006
Production Management: Steffen Jörg
Coverdesign: Stephan Rönigk
Printed and bound by Books on Demand, Norderstedt
Printed in Germany

Table Of Contents

Section A - Indicators

	Issue Date	Page No.
Dial Versus Digital Indicators	Jan-92	2
Test Indicators Vs Dial Indicators	Feb-95	3
Less Is More With One-Rev Indicators	Oct-94	5
One "Special"—To Go	Feb-92	6
Tight, Clean And Dry: Keeping Your Dial Indicator Running Right	Oct-91	8
When Indicators Go Both Ways	Apr-98	9
Electrical Limit Switches: Not New, But Tried And True	Jul-00	11
Going To Extremes	Jan-00	12
Dial Comparators Bridge The Resolution Gap	Feb-00	14
Dial Indicators Are On Every Good Team's Bench	Mar-01	15
Reverse Polarity: Countering Your Clockwise Indicator	Apr-03	17
Indicator Troubleshooting And Repair	Mar-05	18

Section B - Basics Of Measurement

	Issue Date	Page No.
Gaging Accuracy: Getting Ready One Step At A Time	Mar-92	22
Gage Accuracy Runs Hot And Cold	Jul-91	23
What Do You Mean By Accuracy	Apr-91	24
Measuring Versus Gaging	May-96	25
Commonly Asked Questions: Picking The Right Gage And Master	Feb-93	27
What Kind Of Gage Do You Need? A Baker's Dozen Factors To Consider	Dec-95	29
Gaging IDs And ODs	Apr-95	31
Gage Contacts: Get The Point?	May-95	33
A Physical Check-Up For Gages	Oct-96	35
Squeezing More Accuracy From A Gaging Situation	Aug-96	36
The Real Dirt About Gaging	May-91	38
Gage Layout Is Up To The User	Mar-95	39
Stage It To Gage It	Jun-97	40
Fixtures Are A Common Source Of Gaging Error	Aug-91	42
Gaging Accuracy Is Spelled S-W-I-P-E	Apr-92	43
Central Intelligence	Jul-98	44
Take A Stand	Sep-98	46
Perfect Gaging In An Imperfect World	Oct-99	48
Tired Of Bickering Over Part Specs? Standardize The Measurement Process	May-00	49
Starting From Zero	Apr-00	50
It Don't Mean A Thing If It's Got That Spring	Oct-00	52
The Enemies List: Nine Causes Of Error In Precision Gaging	Jan-06	53
Say Hello To Mr Abbé, Mr Hooke, And Mr Hertz	May-02	54
A Different Differential	Aug-02	56
Another Way To Square It Up	Jan-03	57
Where The Rubber Meets The Road—A Probing Look At Probes	Nov-03	59
Dimensional Collateral: Do Two Sines Equal A Cosine?	Oct-03	61

Section B - Basics Of Measurement *(continued)*

	Issue Date	Page No.
Getting My Stars Aligned	Aug-03	63
Plate Gages 4 To 5, You'll Get It Right	May-04	65
TIR Versus Concentricity For Coaxiality	Apr-05	66
Speeding And Gaging Don't Mix	Jun-05	68

Section C - Surface

	Issue Date	Page No.
Start To Finish	May-94	72
R_x For R_a Measurements	Jul-94	73
Look Into My Stylii: Care Of Surface Finish Gage Contacts	Mar-98	75
Measuring Roughness With Buttons And Donuts	Nov-98	77
Surface Texture From R_a To R_z	Nov-02	78
Searching For The Perfect, Identical Waves	Feb-04	80
Hardness Testing And Surface Variation	Jul-05	81

Section D - Electronics

	Issue Date	Page No.
Amplifiers: More Than Just Readout Devices	Dec-94	84
It's An Analog World	Apr-97	85
Three Heads Are Better Than One	Jan-97	87
Electronic Gaging Basics	May-98	89
Gaging By Computer	Dec-98	91
Electronic Height Gages	May-99	92

Section E - Calibration

	Issue Date	Page No.
Calibrating Gages: Your Place Or Mine	Sep-94	96
Mastering For ID's And OD's	Jun-96	98
Control Thy Gages	Nov-97	99
Gaging And Mastering Uncertainty	Jun-94	102
What's So Accurate About Uncertainty?	Dec-04	103
GR&R Measures More Than Just The Gage	Oct-93	104
What's Wrong With This Picture?	Sep-00	106
With Master Rings And Engagements, Size Matters	Jan-04	107

Section F - Holes

	Issue Date	Page No.
Economical Choice Of Bore Gage Depends Upon Your Application	Jun-91	110
Measuring Deep Holes	Nov-93	111
Long Range Bore Gages	Mar-97	113
Checking Bores For Ovality And Taper	Jul-92	114
Measuring Blind Holes And Counterbores	Aug-93	115
A Shallow Bore (This Is Not An Autobiography)	Oct-98	117
Gaging Countersunk And Chamfered Holes	Jul-99	118
Setting Adjustable Bore Gages	Aug-05	119

Section G - SPC

	Issue Date	Page No.
Gaging For SPC: Keeping It Simple	Dec-92	124
Bedrock SQC	Feb-96	125
More On Gaging Statistics	Mar-96	127
Assessing Gage Stability	Mar-99	129

Section H - Machine Calibration

	Issue Date	Page No.
Calibrating Machines For Quality	Mar-93	132
'Round And 'Round She Goes	Jun-95	133
Is Your Machine Square?	Oct-95	134
Evaluating Machine Tools With Lasers	Apr-96	136
Let's Level	Nov-91	137
Using Differential Levels	Nov-94	138
Use A Straightedge To Assess Machine Tool Accuracy	Aug-95	139

Section I - Gage Blocks

	Issue Date	Page No.
Getting Ready For The Microinch Revolution	Feb-94	142
The Mechanics Of Millionth Measurements	Apr-94	143
Don't Wring Your Hands, Wring Your Gage Blocks	May-92	145
Clean It Up And Wring It Dry	Jun-92	146
Working With Your Working Gage Blocks	Mar-02	147
Gage Block Verification	Sep-03	148
New Gage Block Standard—Location Is Everything	Jun-04	150
Formulating Sources Of Error In Your Form Measuring System	Sep-04	152

Section J - Geometry

	Issue Date	Page No.
The Shape Of Things To Come	Jan-93	156
Roundness Gaging—Approximately	Jan-94	157
Circular Geometry Gaging Means More Than Roundness	Sep-97	159
Geometry Gaging Part I	Jul-97	161
Geometry Gaging Part II: Four Methods Of Measuring Out-Of-Roundness	Aug-97	163
Air Rings, CMMs And Supermikes	Dec-93	164

Section K - Air

	Issue Date	Page No.
You Won't Err With Air	Sep-01	168
Flexibility Of Air Gaging	Oct-92	169
Air Is Free, But Not Carefree	Sep-92	171
Air Gaging For Itty-Bitty Holes	Jan-98	172
Choosing The Right Air Gage	May-97	174
Checking For Centralization And Balance Errors	Aug-99	175
The 3 D's Of Straightness Plugs	Sep-02	177
Sliced Bread And The Limits Of Air Gaging	Dec-02	178
Advances In Air Gaging	May-03	180

Section L - Automatic

	Issue Date	Page No.
Process Control Gaging	Nov-96	184
Automated Gaging	Feb-97	185
Machine Compensation	Nov-99	187
Semi-Automatics—The In-Between Gages	Feb-02	188

Section M - Hand Tools

	Issue Date	Page No.
OD Gaging Can Be A Snap	Jun-93	192
Using Adjustable Snap Gages	Jul-93	193
Inspecting Multiple Diameters	Oct-97	195
Calipers: Ideal For Measurement On The Go	Dec-99	196
Micrometers: Measuring Under The Influence	Mar-00	198
"Stylin" With Your Micrometer	Dec-00	199
Stacking Up For Big ID's	Jul-02	201
Micrometer Accuracy: Drunken Threads And Slip-Sticks	Jun-03	202
Evaluating Gaging For The Shop Floor	Jul-04	203

Section N - Applications

	Issue Date	Page No.
Gaging The "Oddball" Application	Jan-95	208
Can't Measure It? Try A Caliper Gage	May-93	209
The Versatile Surface Plate	Jul-95	211
The Plane Truth About Flatness	Sep-95	213
Bar Talk: What's Your Sine?	Nov-95	214
Measuring Tapered Parts	Jan-96	216
Depth Gages	Dec-96	217
Got A Match?	Nov-92	219
In The Groove	Dec-97	221
"Squeeze" Gaging	Dec-91	222
Using And Measuring Precision Balls	Aug-94	224
Temperature Compensation	Aug-98	225
Inspecting Tapers—Part 1: Certifying The Master	Jan-99	227
Inspecting Tapers—Part 2: Toolholder Gaging	Apr-99	228
Gaging Distance Between Hole Centers	Feb-98	230
Gaging "Relational" Dimensions	Feb-99	232
Beyond The Height Gage And Surface Plate	Jun-00	233
Deep Thinking About Depth Gages And Their Evolution	Aug-00	235
3, 2, 1, Contact Measuring Thickness	Nov-00	236
The "Issues" With Height Gages	Jan-02	237
An Inside Look At Special Diameters	Jul-03	238
Holes Big Enough To Fall Into	Aug-01	240
Improving Height Gage Results	Mar-04	241
Air Gaging Styrofoam	Apr-04	243

Introduction

Dimensional gaging is both a science and an art. As a science, it must be guided by sound principles that reflect knowledge of matter and of mathematics. Dimensional gaging must also be practiced following sound methods and techniques, with an earnest yet disciplined regard for what the results communicate about the size and shape of an object. Thus, this process reveals the inner character of the person doing it. This makes dimensional gaging a form of self-expression, a work of art—and a rather lively and fascinating art form indeed.

Dimensional gaging, then, requires that you know what you are doing, that you do it in the right way, and even that you do it in the proper spirit. It exercises the problem-solving ability, the inventiveness and the creativity of anyone who takes it seriously. It calls for skill and style, but always in the service of yielding the most reliable measurement data.

The columns collected in this book were intended to help readers meet these requirements; that is, to help them understand the principles and improve their techniques. The columns were designed to make readers both better scientists and better "artists," if you will.

True to their subject matter, these columns present good science and good art. They are both right and well written. If dimensional gaging is vital and lively, then any writing about it had better not be uninformed or dull, although I'm not sure which of these faults would be the more deadly. Fortunately, the writers whose work appears in this collection suffer from neither. This book is both useful and fun to read, as it truly should be.

The fact that two authors contributed the columns attests to the dual nature of dimensional gaging as science and art. Both authors shared this viewpoint because both were masters of these two aspects.

The idea for a monthly column about dimensional gaging originated in 1991. Ken Gettelman was editor of MODERN MACHINE SHOP when Federal Products suggested the possibility. At the time, a major shift was taking place in the manufacturing industry. Many companies were moving responsibility for inspection and quality control from separate QC departments to the shop floor. At the same time, workpieces were becoming more complex, while tolerances were becoming tighter. Competition among manufacturing was becoming stronger, a trend that intensified with the entry of overseas companies into the U.S. market and the emergence of a global economy.

As a leading producer of dimensional gaging devices, Federal Products was particularly quick to recognize this shift. The company saw that it needed to take an active role in educating this new class of dimensional gaging practitioners. MODERN MACHINE SHOP, with its focus on metalworking manufacturing from a shopfloor perspective, was the most appropriate place for the appearance of an educational column about dimensional gaging.

Ken, who was a keen and insightful observer of industry trends, could see the value of the proposed series of columns. He also understood that accepting them from an author representing only one company from the vendor community was not without risk. The columnist would have to adhere to the magazine's strict stand on the objectivity and non-commercial nature of contributed material. With assurance that Federal Products' aim was purely to instruct and inform, Ken agreed that the plan should go ahead. After the details were settled, *Quality Gaging Tips* made its first appearance in the magazine in April 1991.

The original contributor was Jim McCusker, a talented and dedicated expert in dimensional gaging who had been with Federal Products since 1979. Jim was respected as an authority on the subject and was known throughout the industry. A 1979 graduate of Providence College with a degree in Marketing Management, Jim received a Master of Business Administration degree in 1987, graduating *magna cum laude*. Jim's reputation and integrity did much to establish the credibility of *Quality Gaging Tips* and to win a wide readership with the MMS audience. He set the tone and scope of the columns, making the combination of reliable technical content with a clear, compelling writing style quite popular with MMS readers.

Sadly, Jim passed away suddenly and unexpectedly in August 1993, at the young age of 36. His passing was a loss not only to his employer, but also to the entire industry. With deep respect for the tradition that Jim's contribution to the column series represented, George J. Schuetz was asked to contribute in Jim's place. Without interrupting the appearance of the column in print, George's first contribution was published in November 1993. He has been providing the column ever since, making him one of the longest-running columnists in MMS's history.

George's background is impressive. He joined Federal Products in 1976 and has worked in the application areas for mechanical and digital indicators, mechanical gages, air tooling, electronic products, special gaging designs, surface finishing and geometry gaging. He has worked with many customers to solve specific gaging problems. Over the years, he as lectured at various colleges, meetings of the American Society of Mechanical Engineers and conferences sponsored by the National Council of Standards Laboratories, and given numerous technical presentations at trade shows and industry events. Besides his monthly entries in *Quality Gaging Tips*, he has written many articles for various trade journals and magazines. George has an AS Degree in Electrical Engineering from Hartford State Technical College, and a BA in Marketing/Management from Rhode Island College. He is an active member of the American Measuring Tool Manufacturers' Association, having served in a number of positions, including President.

During George's tenure as the *Quality Gaging Tips* columnist, Federal Products completed a merger with Carl Mahr Holding GmbH, a dimensional metrology instrument manufacturer headquartered in Gottingen, Germany, and became Mahr Federal Inc. Currently, George is responsible for precision gage product management and North American channel management for Mahr Federal.

George has not only fulfilled the legacy of his predecessor, but he has also expanded the column's scope and broadened its appeal by keeping the topics in tune with the changing needs of his audience. He has faithfully maintained the tradition of producing material that is always engaging and easy to read, never condescending or arcane.

Although dimensional gaging is a dynamic field, with new developments in measurement devices and techniques, the fundamental principles and basic techniques are timeless. Thus, a collection of *Quality Gaging Tips* has lasting value. This book incorporates 143 of these columns arranged by topic and complete with graphics that may not have appeared in the magazine. This arrangement makes it easy to find specific columns for help in solving problems or refining new techniques. Both new users and experienced personnel will find this book useful as a reference and guide. However, the book also lends itself to fruitful browsing. I invite you to scan the contents page and turn to any column that catches your attention. I'm sure you will find the contents valuable and interesting. Reading this book will likely reinforce a grasp of dimensional gaging science as well as renew your respect and admiration for the art of it.

Mark Albert
Editor-in-Chief
MODERN MACHINE SHOP
February 2006

Section A
Indicators

Dial Versus Digital Indicators

When digital electronic indicators were introduced in the early 1980's, some observers expected them to blow mechanical dial indicators out of the water. But in spite of the clear superiority of electronic indicators for use in statistical process control and data collection systems, mechanical indicators retain other advantages, and they are still frequently specified by many sophisticated users. Neither type is "better" than the other: the choice depends upon the application and the user's personal preference.

The clearest advantage of electronic indicators is in their use for data collection in process control. Electronic indicators can output measurements directly to printers or SPC programs with no operator errors in reading or recording. The operator only has to position the workpiece and press a button: he needn't even read the measurement. With dial indicators, the operator must interpret the pointer's position to read the measurement, then he must record it—generally by hand—and finally the data must be keyed into a computer. That makes three steps during which errors can and frequently do occur. In any situation, where data must be entered into a computer system, digital indicators are the only way to go. Of course, the user pays for this convenience: digital indicators usually cost significantly more than their mechanical counterparts.

Aside from the cost benefit, there is a great deal to be said for mechanical dial indicators. In many ways, the human brain is like an analog device, and it can often gather more information, more quickly, from an analog readout.

Remember how soundly the marketplace rejected digital speedometers in cars? When a measurement need only fall within a certain tolerance range, analog dials are often quicker and easier to read. An experienced gage operator can simply see whether the pointer is within tolerances without taking the time to actually read and interpret the numbers on the dial.

I have seen QC inspectors make consistently accurate go/no-go readings with dial indicators even before the pointer has stopped moving! They can tell at a glance approximately where the pointer will stop, and in many applications, that is close enough. Electronic indicators don't give you the option of approximating. When a digital device is flickering between six and seven, all of the elements in an LCD display may be lit, appearing as an eight. See accompanying illustration.

Skilled operators can "split grads" with dial indicators, i.e., resolve the pointer's position to an accuracy of about one-fifth of the gage's stated minimum graduation value. And analog dials enable the machinist to observe the direction his process is headed. If reading #1 measures $1/5$ of a grad over zero, reading #2 is precisely zero, and reading #3 is $1/5$ of a

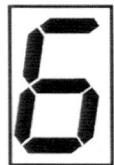 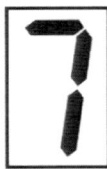 

While electronic indicators are superior for use in data collection and SPC applications, "threshold" readings can sometimes cause problems. When a digital indicator flickers between six and seven, for example, it is very easy to misread the digit as an eight.

grad below zero, the user may be able to draw valuable conclusions about the condition of his tool. In other words, dials can provide more information than simply the dimensional measurement. A digital readout would read zero in all three cases, depriving the user of this additional information. On the other hand, for statistical process control purposes, it is necessary to eliminate all such interpretive data, which again recommends the digital solution.

A common, serious problem among users of dial indicators is the failure to notice when the pointer makes a full revolution or two. Parts that are grossly out of tolerance may appear to be within tolerances to an inattentive operator. In contrast, digital indicators never come "back to zero," eliminating this problem entirely. Furthermore, all digital indicators can be made to signal out-of-tolerance dimensions.

Many electronic indicators have some form of supplemental analog display. These electronic emulations of analog performance serve to eliminate some of the cognitive disadvantages of digital displays and make electronic indicators "user-friendly."

In spite of initial doubts, electronic indicators have proven to be highly reliable in the shop floor environment. Most have only a single moving part, so they may require less frequent cleaning than their mechanical cousins. With proper care, however, dial indicators last forever, and they never need batteries.

Finally, there are somewhat broader ranges of accessories for mechanical indicators, and they are more readily customized for special applications—the subject of another column.

Test Indicators Vs. Dial Indicators

Test indicators are pretty distinct from dial indicators. The immediately obvious difference is that test indicators have lever-type contacts, while dial indicators have plunger-type contacts. Test indicators are also smaller and lighter than dial indicators. In general, the two tools are used in different applications, although there are areas of overlap, where either tool can do the job.

Dial indicators excel at repetitive, comparative measurements: when mounted in a fixture gage, the dial indicator's straight, vertical motion ensures that the contact always lands in the same place, relative to the fixture. This means that the indicator must be oriented vertically to the feature being measured, but for rapid quality inspection of part dimensions, a fixture gage equipped with a dial indicator is unbeatable in most circumstances.

Test indicators excel at consistency measurements, as opposed to comparative ones. They are used most often to explore relatively broad part surfaces in either one or two dimensions—for example, measuring variations in height, flatness, or roundness. Test indicators are often used in combination with a height stand and a surface plate, and either the workpiece or the stand can be moved around freely on the plate. When combined with a V-block or a pair of centers, test indicators can be used to test for roundness or runout on cylindrical parts. The angular motion of the test indicator's lever allows the contact to ride easily over irregularities on part surfaces. This capability is lacking in dial indicators, because the vertical-action plunger may resist responding to surface irregularities pushing "sideways" against the contact.

This ability to ride over irregular surfaces also makes test indicators well suited for use in machine setups, particularly on lathes. The indicator is held by an articulated test stand, usually mounted right on the machine. The operator brings the indicator into rough contact with the chucked blank, then turns the spindle to obtain a very quick reading on runout. No mastering is required when checking roundness, runout, or flatness. You simply bring the indicator close to the part surface, push down on the lever to make contact with the part, and rotate the indicator's bezel to zero. It's far quicker than the typical setup for a dial indicator.

Test indicators can be oriented more flexibly relative to the workpiece than dial indicators, at a wide range of approach angles. The narrow lever and very small contact ball also fit readily into many places that dial indicators cannot reach, except with special attachments. On the other hand, test indicators cannot measure the depth of holes as dial indicators can. Neither are they well suited for use in most fixture gaging applications, nor in ID and OD gages, bore gages, thickness and height gages. These are all standard, no-questions-asked applications for dial indicators.

Spring force is much lower on test indicators, which may be desirable when measuring deformable materials. Test indicators are smaller and lighter than dial indicators, and these factors may be an issue in some fixtures. The dial itself on a test indicator is small, compared to those on dial indicators, so visibility is not as good. As with dial indicators, however, custom dial faces are available for test indicators for special applications.

Test indicators have generally higher resolution, but a shorter range of measurement, than dial indicators, although both factors overlap at the ends of the scales. Typical resolution (least grad) for test indicators is .0001" to .00005"; for dial indicators, it is .001" to .0001". Dial indicators usually have a total measurement range of at least .025", and .250" is also considered a standard figure, while long-travel units allow measurements out to several inches. The measurement range of test indicators is considerably shorter—usually between .008" and .030".

A few more differences to note: test indicators allow just a single revolution of the pointer around the dial (which is part of the reason for their relatively limited range of measurement), while most dial indicators allow 2.5 revolutions (or more, with revolution counters). Test indicator pointers always travel clockwise, and the dials are continuous reading—i.e., the numbers keep ascending until the dial comes back to zero. In contrast, most dial indicators are available with either clockwise or counterclockwise motion, and offer a choice of continuous reading dials or "balanced" ones, with negative values on one side of zero, and positive values on the other.

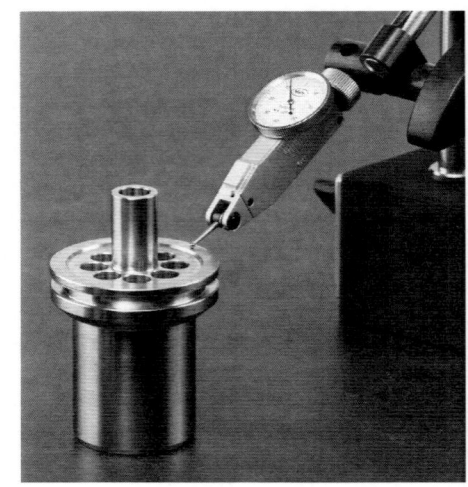

Test indicators are extremely useful little items that are sometimes overlooked in favor of the more familiar dial indicators. When choosing between the two, it's a matter of comparing their relative strengths and weaknesses in light of the requirements of the application.

Less Is More With One-Rev Indicators

Most dial indicators have a total measurement range of 2.5 revolutions of the needle, as per AGD (American Gage Design) specifications. Indicators that allow the needle to go around only once are comparatively rare, but because of some distinct advantages for shop-floor inspection applications, I expect their increased acceptance in the near future. They may even displace AGD indicators as the most popular type.

This added range in the traditional indicator was useful many years ago when machine tool accuracy demanded a broad measurement range to help machinists "creep up" on a specification. Nowadays, though, gaging suppliers recommend that an indicator be chosen so that the tolerance range for the parts being measured should cover between one-tenth and one-quarter of a single needle revolution. This provides a large enough tolerance zone to read easily, and leaves more than enough area on the dial to see what's out-of-tolerance. It's a rare occurrence when anyone actually bothers to read a gage if a part is more than a half-revolution out of tolerance.

Two and a half revolutions are simply unnecessary for most comparative gaging applications, and sometimes they're a real liability. Considering how quickly the needle swings on an indicator, it's not surprising that machinists occasionally miss a revolution. As shown in Figure 1, a measurement that is a full revolution out of tolerance can appear to be exactly on spec to an operator who is distracted—or poorly trained, poorly supervised, or hurried. Errors may occur through simple inattentiveness, or through an absolute misunderstanding of how to set up and master the gage.

There is at least one documented instance in the aircraft industry where an entire run of oversize parts passed through inspection, and was assembled into components, which were subsequently installed in subassemblies. It is not documented what happened to the machinist/operator or his supervisor when this costly error was discovered. But the situation has surely been repeated in other companies and other industries.

Aside from better operator training, there are a couple of ways to minimize this problem. One solution is to use indicators with revolution counters—little accessory dials that

Fig. 1

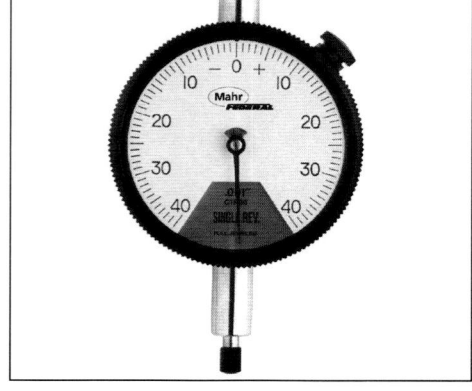

Fig. 2

show the operator how many times the main needle has swung around. The problem here is that an operator can still ignore or misinterpret the tiny rev counter. Indicators with rev counters are actually better suited to long-travel applications, such as where the spindle must clear an obstruction on the part, or where measurements are absolute, as opposed to comparative.

A better solution is the one-revolution indicator, as shown in Figure 2. These have the same range per revolution as comparable AGD-spec indicators. (In other words, the needle moves the same distance for a given amount of spindle travel.) They also have the same amount of spindle travel, so they can be used on parts that are just as far out-of-spec. But in a one-rev indicator, the needle stops moving after one complete revolution (actually, a bit less: usually 340° to 350°), coming to rest in a "dead zone," even if the spindle keeps traveling.

One-rev indicators always have balanced dials, with zero falling between an "over" side and an "under" side: they don't allow continuous clockwise or counterclockwise readings. The needle of a one-rev indicator cannot come back to zero. It can't even travel a full 180° from zero, so there is virtually no way that "over" can be confused with "under," or that out-of-tolerance can be construed as in-tolerance. For less sophisticated operators, or anyone who performs quick, repetitive part inspections in a production environment, one-rev indicators can eliminate a major cause of misreadings. Applications include snap gages, bore gages, and many other comparative–type inspection gages.

The only liability of one-rev indicators, relative to AGD-spec indicators, is that more sophisticated users will give up the capability to measure broader ranges of part variation—for example, when an operator needs to know just how far out of tolerance a part is. Conventional multi-rev indicators still have an important role to play where greater measuring flexibility is required.

One "Special"—To Go

Gage users frequently ask about custom-engineered dial indicators. They are often pleased to discover that some manufacturers specialize in "specials" and make them an important part of their business. Better yet, re-engineering is rarely necessary. Minor changes suffice to make stock indicators appropriate for most unconventional uses. "Specials" can make indicator gaging easier, quicker or more accurate for many applications.

In many gage setups, the indicator's sensitive contact moves in some proportion to a change in workpiece dimension. In the diameter gage shown, two reference contacts provide the benefit of self-centering positioning, but this means that the indicator's sensitive contact is measuring a perpendicular to a chord, not the diameter itself. There is a fixed ratio of 5:4 between the workpiece diameter and this perpendicular. The indicator has a special-ratio face that reads five units for every four units of movement at the sensitive contact, allowing for easy, "direct" readings. If the indicator has a stock face, the user would have to multiply all of his readings by 1.25 to find the diameter. The special-ratio dial saves time for the user, eliminates a source of potential error, and it requires no re-engineering—just a custom-printed face.

Stock indicators can be used to check gross dimensions—as in rough metal castings.

An indicator whose total range is 0.100 inch can measure dimensions up to 1 inch by using a 10:1 lever at the sensitive contact. A special face converts 0.001 inch of movement at the indicator's sensitive contact into a reading of 0.01 inch.

If your shop works to both metric and inch specs, a combination dial, showing both types of units, may simplify setup and reduce purchasing requirements.

Faces can be shaded for different purposes. In "spotlight gaging," the green-shaded area of the dial indicates that the workpiece is within tolerance; the yellow areas warn the user that he is approaching tolerance limits, and the red area means he is out of tolerance. Shaded dials can also be used to quickly sort parts by size, using color-coded bins to coordinate with zones on the dial. For go/no-go gaging, areas of the dial can be masked so that the pointer cannot be seen at all. If the pointer is not visible, the part is no good.

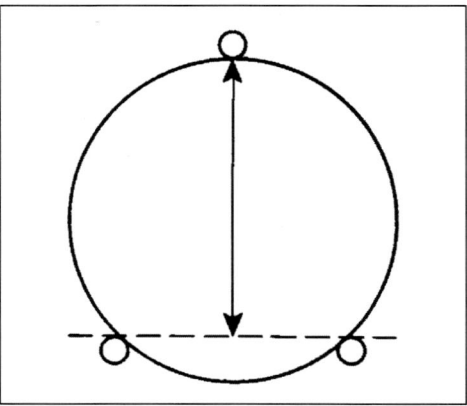

Gages that automatically center the workpiece can greatly speed up the operation. The paired reference contacts quickly locate the part relative to the sensitive contact. However, as the arrows indicate, a chord is measured rather than a diameter. So, the dial indicator used has a special ratio and dial that shows the true diameter.

Dial indicators can be used to measure non-dimensional units. For example, where a relationship exists between temperature and the deflection of a material, a special dial may be used to measure in degrees of temperature. The penetration of a probe into a metal sample can be converted on the dial to read in units of hardness (e.g., Rockwell scale). Special dials can be designed to read in whatever units the industry or application requires. Other examples include: foot-pounds of torque, degrees of angle, pounds of impact force, spring force or cable tension, compressibility, and even diopters.

Beyond special dials, indicators can be modified for extreme environments. "Wet-proof" indicators incorporate rubber O-rings and boots, heavy-duty caps and double bezels for use in dirty or wet environments. Indicators engineered from carefully selected materials can be used at high temperatures—up to about 600°F.

In most indicators, the pointer moves clockwise as the plunger is depressed, and most dial faces have a plus sign (+) to the right of zero and a minus sign (-) to the left. Many

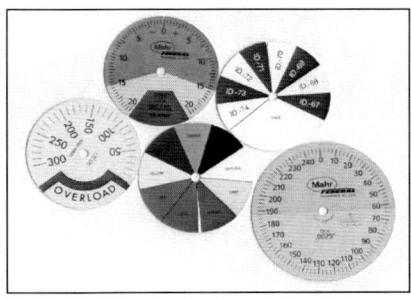

Dial faces can be shaded for many purposes. In addition to showing tolerance, they can help sort parts by size, count pointer revolutions, or read in non-dimensional units such as temperature or hardness.

Section A: Indicators 7

depth gages and bore gages, however, put the plus sign (+) on the left, so that deeper holes or larger IDs read as "bigger" as the spindle moves farther out of the case. Some users, however, like to see the plus sign (+) to the right of O, even when they are measuring holes or bores. To accommodate this preference, indicators are available with reverse movements, in which the pointer moves counterclockwise as the sensitive contact is depressed. In other words a "bigger" hole makes the pointer move into the plus (+) range, to the right of the zero.

In "push-down" movements, the spindle is at rest in the up, or retracted, position. To operate, the user presses the spindle down against the workpiece. This is useful to ensure proper location of the contact on parts with difficult contours, or to avoid interference with the indicator spindle when placing the workpiece in a fixture.

Long-range indicators have long spindle movement and supplementary dials to count pointer revolutions. These can be useful to provide clearance over obstructions, or to make absolute—as opposed to comparative—measurements of as much as six inches.

So when dial indicator gaging is difficult, time consuming or confusing, look beyond the manufacturer's catalog and give him a call. Chances are, there is a "Special" to make the job easier.

Tight, Clean And Dry:
Keeping Your Dial Indicator Running Right

Having found the correct indicator for your application, here are a number of tips to keep it working smoothly and accurately.

First, mount your dial indicator correctly. The ideal method is to mount it from the back, using one of the optionally available lug or rack-type backs available from most suppliers. Mounting by the case or the stem is less desirable, because these components are part of the mechanism. Do not allow a setscrew to bear directly on the stem—the stem will deform, interfering with the movement of the spindle in its bearings. If the indicator must be mounted by the stem, it is essential to use a split bushing or a collet to distribute the clamping force evenly. It should go without saying that the indicator must be mounted securely to the fixture, with no wobble or play.

Worn or loose contacts can also cause false readings, so it is essential to inspect them frequently for wear and tightness. The contact should be screwed onto the spindle "fingertip tight"—just tight enough so it does not loosen up during use. Do not use pliers or a wrench—too much torque will distort the spindle, causing the mechanism to bind.

Replace contacts as soon as wear is detectable. If wear is rapid, consider changing to a harder material.

Hardened steel contacts wear quickly when used against rough or abrasive surfaces and may also be affected by corrosive agents in the work environment. Chromium steel contacts offer better corrosion resistance, but are only marginally tougher than hardened steel. Tungsten carbide or diamond contacts are often the most cost-effective, even though they are the most expensive. They resist wear much longer, thus reducing the need for replacement parts and labor. More importantly, less wear means the indicator will produce fewer false readings.

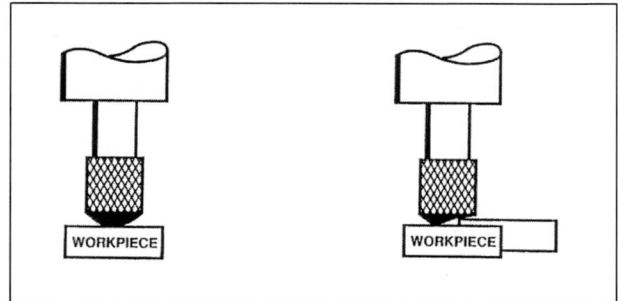
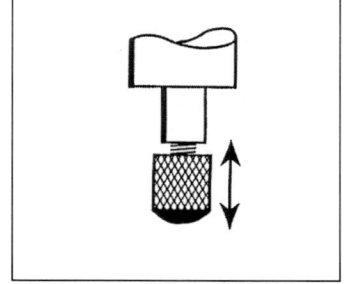

Workpieces should always be fully positioned under the measuring contact. When contact point is worn, an improperly positioned workpiece can cause significant error.

Loose contact points are one of the most frequent causes of measurement error.

If the indicator has been sitting idle for a while, the spindle may stick. Do not oil it. Work it in and out a few times by hand: chances are it will free up. Oil acts like a magnet for dust. Every time the spindle retracts into the case, it will pull contaminants into the indicator's precision movement. The oil, itself, will also harden with time, causing even more sticking. Too often, the problem of sticky oil is improperly treated by adding more oil, until the movement gets loaded with a gummy mess.

The only part of an indicator that should ever be lubricated is the jeweled movement. Manufacturers typically use the point of a pin to apply a minuscule amount of watch-grade oil at this location (a drop of oil from a can would be about 20 times too much). Only individuals trained in the proper methods should open up the case to oil the jeweled movement. Dial indicators that have been cared for properly will rarely require this.

Use a soft, lint-free cloth to remove dirt and oil from the spindle. Clean the crystal with soapy water, benzene, or a soft eraser. Replace scratched crystals and illegible faces. If the indicator looks like it is in poor shape, chances are it will be abused even more. If you keep it clean, it will be treated like the precision instrument it is. And that will mean years—perhaps decades—of accurate measurements and trouble-free use.

When Indicators Go Both Ways

Measuring and gaging are two fairly distinct forms of dimensional inspection. Measuring is a direct-reading process, in which the instrument incorporates a continuous scale of units, against which the part is compared directly. Examples of measuring instruments include steel rules, Vernier calipers, and micrometers.

Gaging is an indirect-reading process, in which the instrument is first set to a standard or master, then the part is compared to that setting (usually defined as zero). This is usually done because the measuring capacity of the gage (i.e., the size of the workpiece it can accommodate) is greater than the measuring capacity of the indicating device (i.e., its range). Gaging instruments include snap gages, bore gages, height and ID/OD comparators, and many others.

There is also a class of instruments that go both ways, blurring the distinction between gaging and measuring. (Hey Gage, take a walk on the wild side...) These are instruments with indicating devices whose range is equal to or greater than the capacity of the gage. Instruments with long-range indicators can perform both direct and indirect measurements.

If zero is set at the reference surface on which the workpiece rests, the indicating unit will *measure* the part directly and display its actual dimension on its scale. Alternately, zero may be set against a master, in which case the instrument becomes an indirect-reading or comparative gage, and displays part size as deviation (plus or minus) from a pre-set value.

Long-range dial indicators that can be used in these "either-or" applications have existed for decades. Although they are familiar and reliable, long-range dial indicators are not easy to read. Workers often make errors when trying to interpret the rather complicated display, which, in addition to the main dial, includes one or two revolution counters rotating in opposite directions.

A recent development is the long-range electronic indicator, with a digital display that eliminates the readability problem of dial indicators. Long-range digital indicators typically have measurement ranges of $1/2$"/12 mm or 1"/25 mm, and resolution of 50 microinches/0.001 mm. They are invariably equipped with lifting levers so that the contact point may be easily raised and lowered. In comparison, long-range dial indicators have ranges from $1/2$"/12 mm to 3"/75 mm or more, but resolution is rarely better than 0.0001"/0.0025 mm, and sometimes as coarse as 0.010"/0.25 mm.

Long-range indicators are typically used on portable thickness gages or bench height/thickness stands. The portable gage provides flexibility for measuring a variety of parts anywhere in the shop, including parts on machines before, in the midst of, and after major stock removal, and multiple dimensions on the same part. Likewise, a bench gage like that

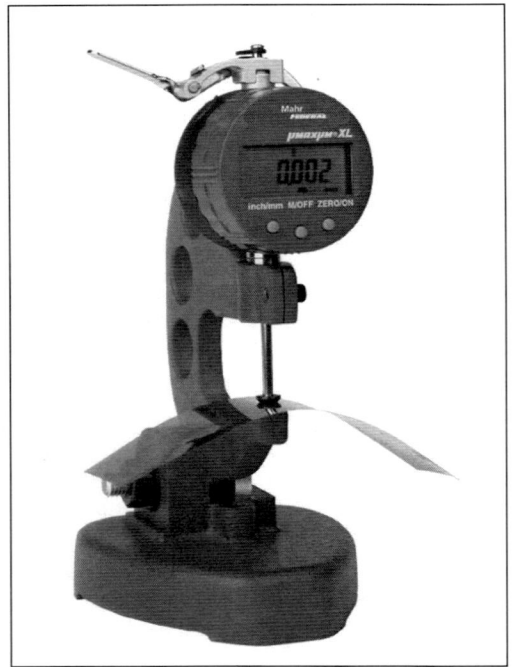

shown in the photo can measure whole families of parts or multiple dimensions on the same part, including thickness, shoulder heights, depths, and outside diameters. Both types of gages offer potential time savings (from the elimination of mastering and setup changes) and cost savings (because one gage does the work of many).

Rules for the use of portable, long-range thickness gages mirror those for snap gages. Make sure the lower or fixed reference anvil is held firmly and squarely against the work, and do not allow the weight of the gage to bear on the sensitive contact. Aside from that, most gages equipped with long-range digital indicators are easy to work with, and highly adaptable.

Electrical Limit Switches: Not New, But Tried And True

There is a category of gages that have been around for fifty years, are very inexpensive compared to alternate measuring methods, are fast, reliable, easy to set up, and can work for manual or automatic operations. I'm referring to mechanical gages that incorporate electrical limit switches. These include a number of different instruments that operate essentially as mechanical displacement gages. While they look and act like mechanical dial indicators or comparators, they include electrical contact points that can be adjusted to represent set points within the range of the indicator. Despite their age, these electric gages still provide the most economical means for classifying or controlling part size in many applications.

In a manual mode, when the switching indicator is combined with a light box, they provide a fast and sure way for operators to classify parts and reduce the possibility of misclassification. The other advantage of this type of gaging is in automatic or semi-automatic gaging operations. If the setup is fairly simple, with one or two gaging stations, there is probably no more cost-effective way to classify parts. Such a classifying indicator would be much less expensive than some of the alternatives, such as classifying amplifiers and related electronics.

There are basically three different types of switching gages.

The most basic type does not even include a dial indicator readout. It provides very repeatable switching within the range of the mechanical sensing head. The switching head typically has a high and low limit switch. These can be electrically combined to give the operator or machine a high, low, or good condition signal. The switching contacts on this type of device can be very accurate. In fact, one of the most impressive features of these gages is the discrimination sensitivity of the switching, which is typically 40 microinches, but some can be good to 10 microinches. This performance approaches that of some much more expensive types of classifying gages. The only drawback is that when using sensing heads without an indicator scale, the two limit positions of the gage must be set with the aid of two masters or gage block stacks representing the limit sizes.

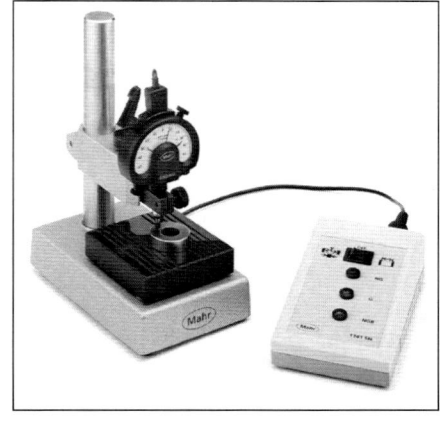

The next style of gage electronic comparator, or electric dial indicator, has a complementary set of electrical contacts, and only requires a zero master for positioning and setting the limit switches. In use, the gage is set so that it is in the center of its mechanical range with the zero master. Then, using adjustment screws, the limits can be set for high and low limit switches. This makes the setup on the gage very fast and simple. It's also simple to adjust for changes in tolerance and adjustment of the part process when required, because it is easy to see the part being measured as the dial hand goes to the limit positions.

The key to any dial indicator with limit switching contacts is the way the contacts are used.

It's very important to put as little electrical stress on the contacts as possible. This means that the current going through the contacts should be minimal. Typically, the voltage level for switching is around 24 volts, but the maximum current should not exceed 100 mA. The purpose of this is to reduce pitting of the switching contacts. When pitting occurs, repeatability errors will increase.

Today, there is a new generation of "on the spot" classifiers. Instead of using the mechanism of a dial indicator and switches, the electronics of a digital dial indicator are given the ability to "look at" the reading, compare it to a set of preset tolerances, and provide an electronic signal representing the out-of-tolerance condition. The advantages of this type of electronic classification include:

- The ability to set the tolerances to the least significant digit of the readout.
- Repeatability of the switching to the least significant digit of the display.
- Operation features found in a digital indicator, such as presets, data output and dynamic measurements.

Sometimes it's easy to put together some of the latest technology to solve your measurement and classification requirements. However, there are some tried and true methods still available that will do an outstanding job and might cost only a fraction of the price you'd otherwise have to pay.

■ ■ ■

Going To Extremes

When You Can't Always Measure In Your Thermal Comfort Zone

When most of us think about measurement environments, what generally comes to mind are pleasant laboratories with temperatures controlled to 68°F/20°C—plus or minus a degree or two. Or in the worst case, we picture a gaging shop with swings of temperature between 65° and 90°F.

Unfortunately, there are also gaging situations where measurements must be made in temperature environments well outside of the inspector's comfort zone. But it does not have to be outside the gages.

1. Oh baby, it's hot in here.

Sometimes a request like this will come across my desk: Our research division is in need of dial indicators that will withstand working conditions of room temperature to 250°F. They will be used in 10,000-hour tests and must remain in the oven for the duration.

The customer was using an oven capable of cycling through a wide range of temperatures to test part performance. A customized indicator with a glass dial and chrome body was supplied to meet the sustained accuracy requirements in the face of frequent thermal cycling.

Here are some of the considerations that go into making a dial indicator worthy of high temperature applications:

Crystal. One of the most obvious considerations is the crystal. Most crystals today are made from plastic blends or alloys that can withstand temperatures up to 170°. To meet the

°F	200	250	300	400	600
°C	79	107	135	190	245
Required Modifications					
Remove All Oil	X	X	X	X	X
Lube With Molycote	X	X	X	X	X
Glass Crystal	X	X	X	X	X
Remove Paint		X	X	X	X
Steel Bezel			X	X	X
Metal Dial				X	X
Steel Back				X	X
Polish Gear Pivots				X	X
Open Rack Bushing				X	X
Steel Hair Spring					X
Steel P.B. Spring					X

250°F temperature requirement, a glass crystal may be substituted.

Bezel. Bezels are typically made of plastic or zinc. For extreme high-temperature gaging, a steel bezel is the most likely choice.

Paints and Coatings. Another consideration is the paint on the indicator and dial. Most paints won't handle this type of temperature. We might consider a special temperature-resistant coating or, for the most extreme applications, none at all.

Lubricant. Technically speaking, indicators are not lubricated. However, a small amount of watch oil is applied to the jewel bearings. At high temperatures, a special Molycote can maintain lubricity when other coatings would break down.

Thermal Expansion Differentials. In dial indicators, there are some very tight clearances. Various materials used in making the indicator will expand and contract at different rates. So it's important for the manufacturer to use materials and clearances that will ensure performance of the gage over the entire operating temperature range of the test.

Clearances between brass bushings and the steel rack, and between the brass gears and the top and bottom steel plates that hold them in place, must all be sized to eliminate the potential binding or slop due to differential expansion or contraction.

The expansion characteristics of the high temperature dial indicator should be provided to the user, making it possible to mathematically adjust measurements to compensate for different rates of expansion in the gage and the part.

Spring Performance. High temperatures may also diminish the force generated by take up and pull back springs within the dial indicators. As a result, better grades of steel and larger diameter wire may be needed to ensure sustained performance of the gage. These types of customizations can accommodate for measuring applications in high temperature environments up to 600°F.

2. Whoa! I'm turning blue.

Sometimes measurements must be taken in environments that are mercilessly cold. The same thought processes are used to determine modifications that will keep the gage, but not the user, from numbing out. But usually lower temperatures are not the biggest problem.

Whichever direction the thermometer is moving in, mechanical indicators still can provide an economical solution to some of the most difficult measurement problems.

■ ■ ■

Dial Comparators Bridge The Resolution Gap

Sometimes it's necessary to measure with very high resolutions approaching 20µ"/0.5µm. Measurements at this end of the spectrum would normally be reserved for an electronic amplifier with high performance electronics. However, there may still be reasons to prefer a mechanical measurement tool for a given job. Primary among them are likely to be budgetary restrictions, or a requirement for portability that would make cabling a hand gage to a transducer less than desirable.

The lowest priced measurement tool is, of course, the dial indicator. But dial indicators are not normally thought of as providing a high-resolution readout. Their high confidence zone is typically up to .0001"/2µm. Manufacturing dial indicators to the highest possible standards will result in only modest improvement of resolution. This is because the substantial number of parts in a dial indicator generate a build-up of tolerances. The high amplification required for ultra-precision measurements tends to magnify these errors, which can show up as degraded accuracy, hysteresis and/or repeatability.

This is why electronic amplifiers have become so important. With only one moving part and the stability of solid-state electronics, the errors are extremely small. Even though an amplifier may have very high magnification, the total error occupies a very small fragment of the instrument's resolution. But again, because of price and portability issues, this may not be the answer you want to hear.

There is another form of mechanical gage that bridges the gap between dial indicator and the electronic amplifier. Known as the dial comparator, it solves the inherent mechanical problems of its close cousin, the geared dial indicator. Here's how.

- Travel of the spindle in a dial comparator is guided by a precision ball guide. This not only eliminates friction, but also provides strong axial stiffness. This assures results in a near one-to-one translation of the motion to indicator movement.

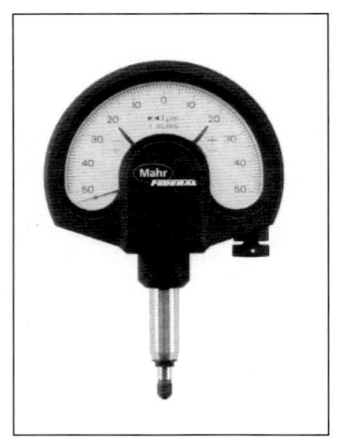

- The ball guide also reduces the long-term damage to the spindle, which can result from side play created when parts are forced from underneath the contact.

- A shock proofing system isolates the spindle from shock created by banging of the gears during rapid movement.

- Many of the gear assemblies used in mechanical amplification have been replaced by simple levers. Fewer components result in a reduced opportunity to magnify errors.

- Jeweled components provide maximum sensitivity and accuracy in the movement of gears and levers.

14   Quality Gaging Tips

- A built-in compression spring provides a constant measuring force over the entire range of the indicator.
- Built-in calibration adjustment allows for fine-tuning out even the most minute errors.

While all of these features assure the highest accuracy, the dial comparator still retains all those features which make dial indicators extremely practical and easy to use. These include:
- A built-in, lockable, fine adjustment control.
- Adjustable tolerance markers.
- Very fast response indicator hand.
- Remote cable retraction of the spindle.

Any analog device has range versus resolution constraints—the dial comparator is no exception. Since it has much higher resolution, the dial comparator also has much less range than the dial indicator. However, this might even be an advantage in some measuring processes. Since these comparators are used in very high tolerance work, limited range is probably not a major factor and might even prevent some measurement errors. Some dial indicators employ a revolution counter. In instances where the operator fails to pay attention to the counter, recorded measurements may be off by a full revolution. A dial comparator is very much like a one-rev indicator. The part has to fall within the measuring range of the indicator or else it shows up as being off-scale. There is no chance of being read incorrectly.

So what's the bottom line here? A dial comparator has five to ten times better resolution and accuracy than a dial indicator. The electronic amplifier has about 10 times better resolution and accuracy than the dial comparator. A good dial indicator can cost about $150. A dial comparator runs anywhere from 30% to 100% more. The cost of an electronic amplifier and probe is almost seven times the price of a good dial indicator. So where range is not a concern, the comparator is your best resolution and accuracy value. If you need high performance on a budget that's a little tight, the dial comparator could be your solution.

Dial Indicators Are On Every Good Team's Bench

Do's And Don'ts For Keeping Them In The Game

Just because dial indicators have been around since the early 1900s, don't expect them to fade away with the last century. This tool's long-term popularity is well earned. Dial indicators offer good resolution at low cost, but that is not the main reason people still use them.

Beyond providing easy-to-read quantitative measurement, dial indicators give users a comparative sense that their parts are in the ballpark. You simply see if the indicator's needle is within tolerance bands or, simpler still, lies within red sections highlighted on the dial. No interpreting is necessary. Every result may not read like a home run, but as long as it's not in the outfield, it scores as a good part.

Dial indicators vary widely in type, size and range. All translate variations (through internal movement of a plunger) into dial readings. Some will indicate dimensional variations as small as 0.00005". So you must handle these sensitive mechanisms with the same devoted care you give to other precision equipment.

To maintain high levels of quality and precision, take heed of the following tips.

By All Means Do:
- Mount dial indicators close to short support columns on test sets or comparators to avoid holding rod deflection.
- Keep the reference surface clean and level, with the test set base clean and seated positively.
- Mount your indicator securely to the fixture or holding device.
- Keep the indicator spindle and point clean, using a soft, lint-free cloth.
- Make sure the indicator hand moves toward the minus side of the dial as workpiece dimension decreases.
- Handle the gage lightly, so it can seat itself on the workpiece.
- Use diamond, tungsten carbide or hard chromium-tipped indicator points whenever it's likely that the contacts will be subject to heavy wear.
- Store your dial indicators in a safe, dry place and cover them to keep the dust and moisture out.
- Test your indicators under gaging conditions at intervals during the operating day. You can do this by gaging a part twice, then comparing its readings to a master part.
- Clean dials with soap and water, benzene or soft eraser. Frequency of cleaning depends on the type of gaging and the contaminants.

For Heaven's Sake Don't:
- Don't subject indicators to harsh, sudden blows. If blows are unavoidable, use a cushioned movement indicator.
- Don't overlook accessories that will make your indicator more efficient, more adaptable and more versatile, e.g., lifting levers, right angle attachments, maximum point hands and weights for measuring compressible materials.
- Don't oil spindle bearings except under special conditions. Then do it sparingly and never use grease.
- Don't tighten contact points or adapters too far against rack spindle, as the strain will cause distortion, make the spindle bind, the mechanism stick or the guidepin loosen or shear off.
- Don't clamp indicator against the stem with a set screw. Too much pressure will make the rack spindle bind, causing the indicator to become sluggish and sticky.
- Don't lock the indicator in position until you've set it carefully under proper gaging tension, that is, at least a quarter turn from its "at rest" position.
- Don't oil an indicator that has been idle for some time. If the spindle sticks, work it in and out by hand until it slides freely on its own bearings.
- Don't drill holes in the back of the case. Chips will get inside and ruin the movement.
- Don't use an indicator that has been dropped or struck until you have it tested thoroughly. Test it on a comparator set or some other supporting device to make sure it's precisely calibrated. Then re-set the indicator in position as precisely as you set it the first time.

- Don't use your dial indicator for anything but what it is intended for—accurate gaging. It's not a jackhammer or paperweight. It won't give good service unless you treat it the same as your other precision instruments.

By following these tips—what to do as well as what not to do—with your dial indicator gages, you can keep them accurate and in the game for a long time.

■ ■ ■

Reverse Polarity: Countering Your Clockwise Indicator

We have looked at the advantages of balanced and continuous dial indicators. There are also measurements that require dial indicators with a counterclockwise dial. The counterclockwise dial can have the same characteristics as the balanced or continuous dial—except that it is reversed. A balanced counterclockwise dial would be indicated by having a plus (+) sign on the left, while a continuous counterclockwise dial would count from zero, counterclockwise around the dial face.

With a counterclockwise dial, the hand travels the same as it would normally, but the scale is reversed. In effect the dial indicator has had its polarity reversed.

The golden rule is that a clockwise dial indicator should be used when the indicator and/or the sensitive contact is on the opposite side of the part from the reference point, while the counterclockwise indicator is used when the indicator and the sensitive contact are on the same side as the reference point.

For example, when an indicator is used on a bench stand, the anvil is the reference point. The part rests on the anvil and the indicator and the sensitive contact touching the part are opposite the reference, as shown in Fig. 1. This application would call for a clockwise dial indicator.

On the other hand, when using a bench thickness gage, as seen in Fig. 2, the reference, the sensitive contact, and the dial indicator are all on the same side of the part—thus the need for a counterclockwise movement. In this bench type of application, when a groove depth gets shorter, it pushes in on the indicator. On a normal indicator this would generate a plus reading. But with a counterclockwise/reversed dial, the indicator reads the correct polarity: i.e., a smaller reading.

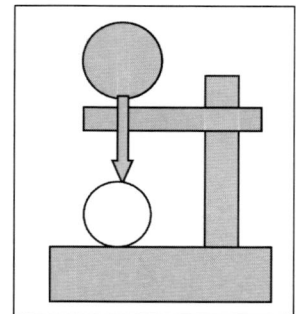

Fig. 1

Let's look at some other examples to demonstrate our golden rule:

First, take a look at the Bench ID/OD comparator in Fig 3. When set up for an inside diameter measurement, we have a reference contact on one side of the part. On the opposite side of the ID is the sensitive contact with the dial indicator on the same side, but on the outside of the part. In this case, as part size increases, the sensitive contact moves away from the reference contact and towards the dial indicator. This requires a plus reading

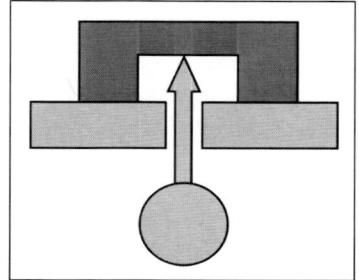

Fig. 2

Section A: Indicators   17

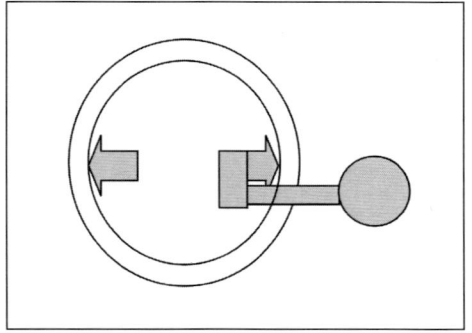

Fig. 3

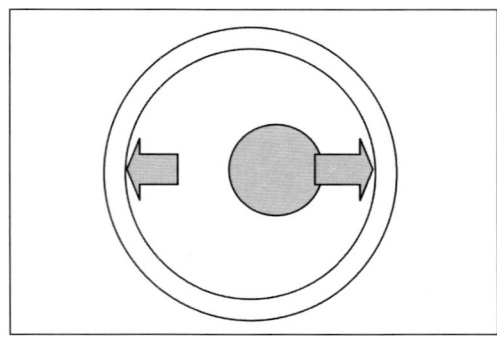

Fig. 4

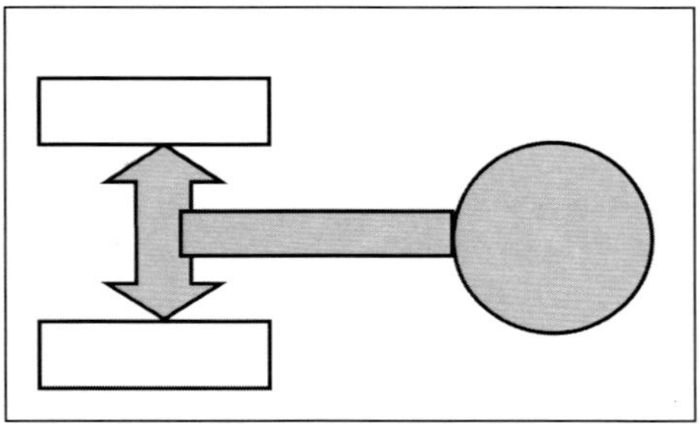

Fig. 5

and thus, a clockwise dial.

Now lets imagine a horizontal bore gage (Fig 4) where the reference contact, the sensitive contact, and the dial indicator are all on the inside of the bore being measured. In this case, as the sensitive contact moves away from the reference contact, the hole is getting bigger, but the indicator would normally display a smaller reading. With a reversed dial this reading would be corrected.

Finally, one of the most common uses of the counterclockwise indicator is in a variable plug gage (Fig 5). At first glance this may appear to contradict our golden rule mentioned above. In this case, the body of the plug is the reference surface. The contacts are on the same side as the reference surface, and so is the dial indicator—but only through its transfer mechanism. Therefore, as the rule states: when all three are on the same side, a reversed indicator should be used.

With mechanical dial indicators, it is a bit difficult to change the dial to/from normal (clockwise) to reverse (counterclockwise) readings. However, with the digital indicators and electronic amplifiers available today—this switching is done with the push of a button.

Indicator Troubleshooting And Repair

Repairing dial indicators is a task usually assigned to a trained gage technician. Since the indicator is the part that amplifies the actual gage displacement to make it a readable value, it is an essential component.

While designed for the shop floor, mechanical indicators and comparators are delicate comparative instruments that need to be handled with care and have their performance checked regularly. However, an indicator may sometimes become faulty or go out of calibration for any number of reasons. In such cases, most users rely on their preferred repair center to fix the indicator and verify its performance.

When time is not an issue, users typically choose between two levels of repair service:

The first is "repair to working" condition. Many repair centers provide a good cleaning, inspect the indicator components, and may re-orient internal parts or incorporate parts from a used indicator to get the gage working. The calibration is then verified and the indicator put back on the application. In many cases, this is a perfectly sound method of indicator repair. However, service warranties may be limited—if provided at all—and the gage should be continuously watched for repeat performance.

The other option, an initially more expensive method but a better value over the life of the indicator, is to have the indicator repaired by the manufacturer or by an a factory authorized repair center. Many manufacturers will repair the indicator to a "like new" condition, often substituting a complete new movement rather then spending time looking for a worn component. And since only original manufactured new components are used, the indicator is recertified and returned with a warranty of service, just as if it were a new indicator.

However, there are also times when things don't go as planned, when an indicator gets stuck at a critical moment and just won't work. At times like these it may seem like the weight of the world is coming down on your head, but don't fear: most indicators manufactured in the past 50 years were meant to be repaired. With a little bit of care and foresight, many of the problems seen in an indicator can be repaired in a pinch—at least, enough to get you through the process, or until a replacement can be brought in to take the place of the suspect indicator.

The symptoms of a sick indicator can vary significantly, but the most common is a stickiness in its travel. One can get so stuck that when the rack is pushed in it simply does not return. And as we have said many times before about gaging, dirt is the probable cause and it may be time for some good housecleaning.

But first take a look at the indicator in general. Is anything missing, or do you see any physical damage? Could the stem be bent and not 90° to the case, a rack bent or a stem pinched from a too tight clamp? If any of these are the case, there is probably not much you can do: it's time for some professional repair or even a new indicator.

If things look OK from the outside, go ahead and remove the indicator's back. Then remove the bezel clamp and the bezel. Some bezels snap on while others have screws that need to be loosened. Once the guts of the indicator are exposed, look for dirt, oil, even floating broken gear teeth, or pieces of a jeweled bearing. In the latter cases, it's again time for a professional.

However, if it just looks dirty or oily, splash alcohol around on the inside of the case until the gears and rack are totally soaked. Work the rack up and down to work out the contamination. Most times, you'll begin to feel a significant difference in the action of the indicator. Repeat the process again and use a little compressed air to blow out the case and dry off the movement. If this routine has eliminated the stickiness, you just might be in business. Put the bezel and clamp back in place and then attach the back, being careful not to over-tighten the back screw, which may distort the case.

But you're not quite ready to put the indicator back on the job. First, run it through its paces to see if it's measuring properly. Mount the indicator in a stand and get some gage blocks that are spaced so that they cover the range over which the indicator will be used. Zero the gage, and start checking the gage blocks. Or, if there is a micro-calibrator available, check the indicator through its full range in both directions. If the indicator is measuring properly (usually to within a grad)—congratulations!—you've saved the day. Put the gage back to work, watch it carefully, and schedule a check-up with the professionals at your earliest convenience. With a good, professional repair, that indicator should be good for another 100,000 cycles.

Section B
Basics Of Measurement

Gaging Accuracy: Getting Ready One Step At A Time

Familiarity may not always breed contempt, but in precision gaging, it can certainly lead to error. It happens when we do what we have done a thousand times before, but do it without thinking. We are in a hurry. We grab a gage and take a measurement without stopping to go through those preliminary checks and procedures we know will assure accurate results. We forget that the methodology of measurement is as important as the gage itself. As a machine operator, you must assume much of the responsibility for gaging accuracy. Whenever a gage has not been in frequent use, make sure you follow these basic steps:

■ Providing the indicator has been checked for calibration, repeatability and free running, look over the way it is clamped to the test set, comparator frame or gage. Any detectable shake or looseness should be corrected.

■ Check for looseness of play in comparator posts, bases, clamping handles, fine adjustment mechanisms and anvils. It is easy, for instance, to rely on the accuracy of a comparator and find afterwards that the reference anvil was not securely clamped down.

■ When using portable or bore gages, be sure to check adjustable or changeable contacts to be sure there is no looseness of play.

■ If gage backstops are to be used and relied on, make sure they are also clamped tight in the proper location.

■ The sensitive contact points on many portable gages and bench comparators are tipped with wear-resisting tungsten carbide, sapphire or diamond inserts. Test these tips to see that they haven't become loose in previous use. Also, examine them under a glass. If they are cracked, chipped or badly scored, their surface conditions may prevent accurate or repeatable readings. They may even scratch the work.

■ If opposing anvils are supposed to be flat or parallel, check them with the wire or ball test. By positioning a precision wire or ball between anvils, you can read parallelism on the indicator simply by moving the wire/ball front to back and side to side.

■ One of the easiest chores to neglect is regular cleaning of indicating gages and bench comparators. Yet, as we have often noted in this column, dirt is the number one enemy of accuracy. Dirt, dust, grit, chips, grease, scum and coolant will interfere with accuracy of gage blocks, indicators, and precision comparators. Clean all such instruments thoroughly at each use. Also, be sure to rustproof exposed iron or steel surfaces.

■ Take the same steps to ensure the reliability of master discs and master rings as you would for gage blocks. Examine them for nicks and scratches and the scars of rough handling. And handle them as you would gage blocks, as well. After all, they are designed to provide equal precision.

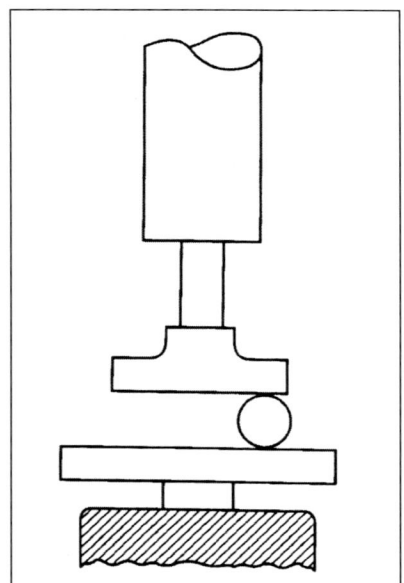

Parallelism between the reference and sensitive anvils can be easily explored with a precision steel ball.

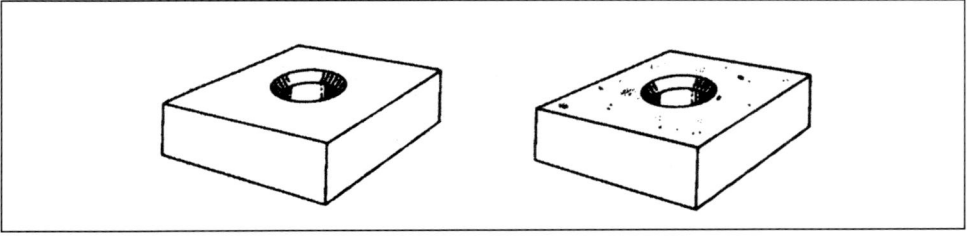

After cleaning a gage block thoroughly, leave it out in the shop environment for 30 minutes, uncovered. Shop moisture and dust can easily add a few ten thousandths to the size of the block.

■ Finally, if you see a sudden shift in your process during the day, these same basic steps should be part of your troubleshooting routine. And, in this situation, don't automatically assume your gage is correct just because it has a calibration sticker. Strange things do happen and you will do well to investigate all possibilities, especially the ones that habit can make us overlook.

■ ■ ■

Gage Accuracy Runs Hot And Cold

"It takes a while to warm up in the morning, but after that, it runs great." I swear I've heard machinists say this of their gages, as if those instruments were like car engines with 50-weight motor oil and cold intake manifolds.

What's really happening, of course, is that the machinist arrives at work, takes his gage and master out of a controlled environment, masters the gage and then gets to work. As he handles it, the gage begins to warm up. Which is not to say that its moving parts move more freely, but instead, that the gage itself expands. Depending on where he keeps his master, and whether or not he re-masters regularly, he will find himself "chasing the reading," possibly for hours, until everything reaches equilibrium.

Thermal effects are among the most pervasive sources of gaging error. Dirt, as a gaging problem, is either there, or it isn't. But everything has a temperature—even properly calibrated gages and masters. The problem arises from the fact that everything else has a temperature too, including the air in the room, the workpiece, the electric lighting overhead, and the operator's fingers. Any one of these "environmental" factors can influence the reading.

Why is temperature such a critical concern? Because most materials expand with heat, and they do so at differing rates. For every 10°F rise in temperature, an inch of steel expands by 60 millionths. "Not to worry," you might say, " I am only working to 'tenths'". But aluminum expands at more than twice that rate, and tungsten carbide at about half. Now, what happens to your reading if you are trying to measure a 2-inch aluminum workpiece with a steel-framed snap gage and tungsten carbide contacts, after the workshop has just warmed up 7 degrees? And by the way, did that workpiece just come off the machine, and how hot is it?

Beats me, too. That's why it's critical to keep the gage, the master, and the workpiece all at the same temperature, and take pains to keep them there.

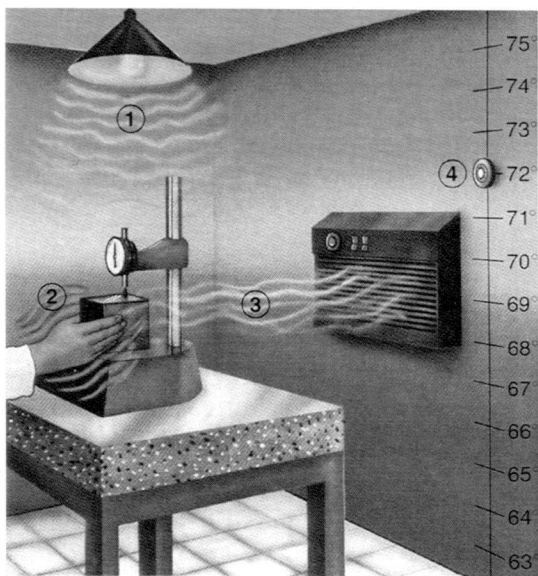

Thermal sources of error are a major cause of gage performance degradation. Typical thermal sources are: (1) radiant heat from illuminating sources; (2) conductive heat (that is, operator touching workpiece); (3) convection and drafts from heating and cooling systems; and (4) room temperature gradients.

That means keeping an eye on many factors. Don't put your master away like some sacred object. Gage and master must be kept together, to ensure that they "grow" in tandem and to permit frequent re-mastering. Workpieces must have sufficient time to reach ambient temperature after machining, or after being moved from room to room. The operator should avoid handling the gage, master and workpiece more than absolutely necessary.

Care must be taken that sources of heat and cold in the room do not intrude on the process. Incandescent lighting, heat and air conditioner ducts, even a shaft of direct sunlight through a window can alter a whole series of measurements. Keep things at the same "altitude" in the room, to avoid the effects of temperature stratification.

As tolerances tighten, additional measures become necessary. Workpieces should be staged on a heat sink beside the gage and should be handled with forceps or gloves. A Plexiglas shield may be required to protect the gage from the operator's breath. (The heat, that is, not the effects of the sardine sandwich he had for lunch.)

For accurate gaging, be aware of possible sources of thermal "contamination" to the measurement process. While it may not be possible to isolate your gaging process in its own perfectly controlled environment, at least take precautions to minimize the effects of temperature variation on your gages, masters and workpieces.

■ ■ ■

What Do You Mean By Accuracy?

How accurate is my gage? How often do you ask yourself that question—checking a dimension on a workpiece, but never fully believing what your gage tells you? You send the piece off and hold your breath while you wait to see if it's accepted or rejected.

Gaging is one of the most critical and least understood operations in machine shops today. Industry can no longer afford yesterday's levels of wastage, and accurate gaging has, therefore, never been more important.

By the same token, the metrology industry has not been consistent in its definitions, but it's important that we agree on certain terms—all of them related to the concept of accuracy—before we can converse intelligently about gaging.

Accuracy, itself, is a nebulous term that incorporates several characteristics of gage performance. Our best definition is that accuracy is the relationship of the workpiece's real dimensions to what the gage says. It's not quantifiable, but it consists of the following quantifiable features.

Precision (also known as repeatability), is the ability of a gage or gaging system to produce the same reading every time the same dimension is measured. A gage can be extremely precise and still highly inaccurate. Picture a bowler who rolls nothing but 7-10 splits time after time. That is precision without accuracy. A gage with poor repeatability will on occasion produce an accurate reading, but it is of little value because you never know when it is right.

Closely related to precision is stability, which is the gage's consistency over a long period of time. The gage may have good precision for the first 15 uses, but how about the first 150? All gages are subject to sources of long-term wear and degradation.

Discrimination is the smallest graduation on the scale, or the last digit of a digital readout. A gage that discriminates to millionths of an inch is of little value if it was built to tolerances of five ten-thousandths.

On analog gages, discrimination is a function of magnification, which is the ratio of the distance traveled by the needle's point to travel at the transducer. A big dial face and a long pointer—high magnification—is an inexpensive way for a manufacturer to provide greater discrimination. This may create the illusion of accuracy, but it isn't necessarily so.

Resolution is the gage's ability to distinguish beyond its discrimination limit. A machinist can generally estimate the pointer's position between two graduations on a dial, but usually not to the resolution of the nearest tenth of a graduation.

Sensitivity is the smallest input that can be detected on the gage. A gage's sensitivity can be higher than its resolution or its precision.

Calibration accuracy measures how closely a gage corresponds to the dimension of known standards throughout its entire measuring range. A gage with good precision may be usable even it its calibration is off, as long as a correction factor is used.

If we could establish these terms in common shop parlance, there would be better agreement about how accurate a gage is.

Measuring Versus Gaging

We often use the terms "gaging" and "measuring" interchangeably, but here, at least, we're going to distinguish between them as different procedures. There are times when gaging is appropriate, and other times when measuring is the way to go. What's the difference?

Measuring is a direct-reading process, in which the inspection instrument consists of (or incorporates) a scale—a continuous series of linear measurement units (i.e., inches or mm), usually from zero up to the maximum capacity of the instrument. The workpiece is compared directly against the scale, and the user counts complete units up from zero, and then fractions of units. The result generated by "measuring" is the actual dimension of the workpiece feature. Examples of measuring instruments include steel rules or scales, vernier calipers, micrometers, and height stands. CMMs might also be placed in this category.

Gages, in contrast, are indirect-reading instruments. The measurement units live not on a scale, but off-site (in a calibration lab somewhere), and a master or other standard object acts as their substitute. The workpiece is directly compared against the master, and only indirectly against the measurement units. The gage thus evaluates not the dimension itself, but the difference between the mastered dimension (i.e., the specification), and the workpiece dimension.

Gages fall into two main categories: "hard," and "variable." "Hard" gages—devices like go/no-go plugs and rings, feeler gages, and non-indicating snap gages—are not conducive to generating numerical results: they usually tell the user only whether the part is good or bad. Variable gages incorporate some principle for sensing and displaying the amount of variation above or below the established dimension. All indicator and comparator gages meet this description, as does air and electronic gaging. The result generated by a variable gage on an accurately sized part is generally 0 (zero), not the dimension. Because of modern industry's need for statistical process control, variable gaging is the norm, and there are few applications for hard gaging.

Variable gaging may be further subdivided into fixed and adjustable gaging. Fixed variable gages, which are designed to inspect a single dimension, include mechanical and air plug gages, and many fixture gages. Adjustable variable gages have a range of adjustment that enables them to be mastered to measure different dimensions. Note that range of adjustability is not synonymous with range of measurement. You can use an adjustable snap gage to inspect a 1" diameter today, and a 3" diameter tomorrow, but it would be impractical to constantly re-master the gage to inspect a mixture of 1" and 3" parts. (This would be no problem for most "measuring" instruments, however.) Almost all indicator gages may be of the adjustable variety.

Because of its relative mechanical simplicity, fixed gaging tends to hold calibration longer, and require less frequent maintenance and mastering. It is often easier and quicker

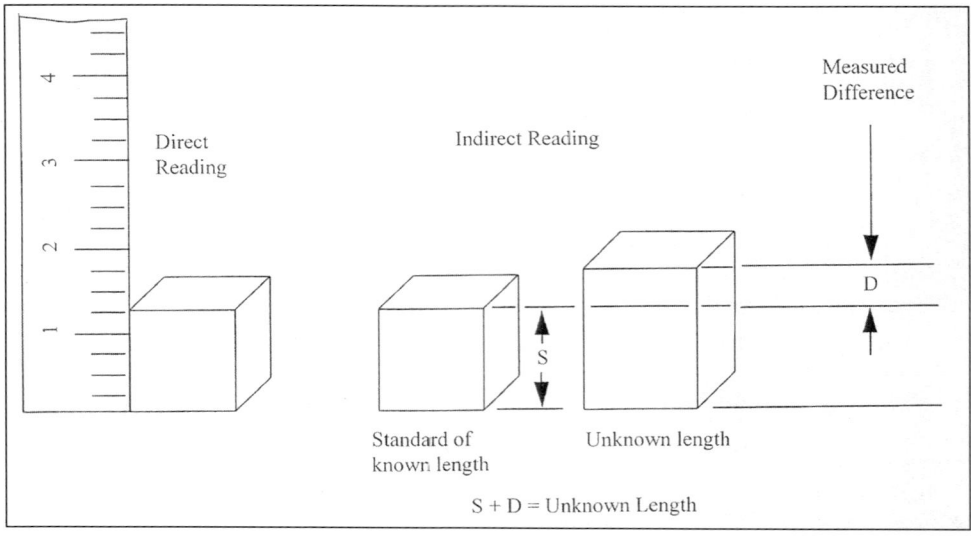

Direct-reading devices (for example, scales and vernier calipers) read the part's actual dimension. Indirect-reading devices (gages) tell the user how far the part deviates from the standard.

to use than adjustable gaging. But it is also inflexible: once a production run has finished, a shop may find it has no further use for a gage designed solely to inspect IDs of 2.2370", ±0.0002".

Where production runs are smaller, or where throughput is not quite so important, adjustable gaging often makes more sense. The range of adjustability allows a gage to be turned toward a new inspection task after the initial one is completed. The adjustable bore gage being used today to measure IDs of 2.2370", ±0.0002" may be used to measure IDs of 1.0875", ±0.0003" next month.

Fixed gaging therefore tends to be economical for inspection tasks that require high throughput, and for production runs that involve many thousands of parts, and that last for months or years. Adjustable gaging tends to be appropriate for shorter production runs, and for smaller shops in general.

Similar issues apply when comparing "gaging" and "measuring." Gaging tends to be faster, both because it is less general-purpose in nature, and because the operator need observe only the last digit or two on a display, rather than count all of the units and decimals up to the present dimension. Because of its generally much shorter range, gaging can also be engineered for higher accuracy (resolution and repeatability) than measuring instruments. For anything resembling a production run, gaging is almost always required. But where single part features must be inspected, measuring devices tend to make more sense. In practice, most shops will find they need some of both types of devices.

Commonly Asked Questions:

Picking The Right Gage And Master

In this job I get asked a lot of questions. In fact, I did some figuring the other day, and estimate, conservatively, that we have probably answered at least 25,000 gaging questions over the past ten years. Some of these questions have been absolutely brilliant. They have pushed me to learn more about my business and our industry, and to grow professionally. Some have even helped me develop new products. Others have been, well... less brilliant. Those asked most often concern picking gages and masters. We have talked about various aspects of this process in previous columns, but I thought it would be well to list the questions and answer them directly. Then, next time someone calls, I can just read the answers!

Without a doubt, the most common question I am asked has to do with selecting a gage: "I've got a bushing with a .750" bore that has to hold ± 0.001 in. What kind of gage should I use?" There are a number of choices: a dial bore gage, an inside micrometer, an air plug, a self-centralizing electronic plug like a Dimentron®, or any one of several other gages. But picking the right gage for your application depends basically on three things: the tolerance you are working with; the volume of components you are producing; and the degree of flexibility you require in the gaging system.

For tolerance, or accuracy, we go back to our ten-to-one rule: if your tolerance is ±0.001 in., you need a gage with an accuracy rating of at least ten times that, or within one tenth (±0.0001 in.). But that's not all there is to it. The gage you pick may also have to pass your own in-house GR&R (Gage Repeatability and Reproducibility) requirements. Just because we, as gage manufacturers, say a gage is accurate to a tenth, doesn't necessarily mean you, as a component manufacturer, will actually get that level of performance from it in the field. GR&R studies are designed to show how repeatable that specified accuracy is when the gage is used by a number of operators, measuring a number of parts in the manufacturing environment. Since this incorporates the whole idea of 'usability,' it makes the process of selecting a gage more complicated. There is no single standard for GR&R studies, but generally, it is a statistical approach to quantifying gage performance under real life conditions. Often this is expressed as the ability to measure within a certain range a certain percent of the time. As "10 percent" is a commonly quoted GR&R number, it should be noted that this is quite different from the traditional ten-to-one rule of thumb. But that's a topic for at least a couple of future columns. For our purposes here, suffice it to say that if passing GR&R is one of your requirements, you should discuss the details with your gage supplier.

Component volume is also of prime importance in picking a gage. How big is the job? How long will it last? How important is it to the shop? This will dictate how much you can spend on a gage or gaging system. Generally speaking, the trade-off here is speed and efficiency for cost and flexibility. You can get a system that will measure several hundred parts an hour, twenty-four hours a day, if that's what you need. But that system is not going to be good at measuring a number of different parts, and it's not going to be inexpensive.

The flip side here is flexibility. It may well be that the decision to buy a gage is based not so much on a specific part, but on overall shop requirements. That may be a different gage from one, which measures a single-sized hole with optimum efficiency. Finally, consider what you intend to do with the reading once you get it. In short, do you need digital output?

After gages, the next most common question concerns masters: what grade and kind to buy. "Do I need XX or XXX, and what's the difference?" The answer here is a bit more direct. There are several classes or grades of masters, depending on accuracy. These are Z, Y, X, XX, and XXX, with Z being the least accurate and the least expensive. Class XX is the most common, with an accuracy rating of ±0.00001 (up to ±0.00005, depending on size). What class you buy is determined, again, by the ten-to-one rule; but based on the gage, not your part. Thus, if your 0.750 in. diameter part has a tolerance of ±0.001 in., pick a gage that is accurate to a tenth (± 0.0001 in.) and a master that is accurate to one-tenth of that, or ten millionths (±0.00001). In this case, that would be a grade XX master.

But now, here's a rub: let's say you have a tolerance of five tenths (± 0.0005 in.) and you are using an air gage with an accuracy of twenty millionths (±0.00002 in.). That is certainly better than ten-to-one for the gage, but what class of master do you use? One that is accurate to two millionths? If so, you've got a problem, because no one makes them. What you do in cases like this is buy a master that is Certified for size. This means it will be accurate to within five millionths (± 0.000005 in.) of the certified size, and will indicate the variation from nominal.

Finally, people continually ask me about chrome plating and carbide. "Why should I pay extra for chrome plating, and when do I need carbide gage blocks or masters?" The answer here is simple, and has to do with the hostility of your gaging environment. Chrome plating protects against corrosion. It is also much more wear resistant than plain steel. So if you have a corrosive or abrasive environment, chrome-plated gages and masters are worth the cost simply because they will last longer.

As for carbide, I generally recommend using blocks and masters of a material similar to the part you are machining, because of thermal expansion. Carbide has a coefficient of thermal expansion about one-third that of steel. If the temperature in the shop changes—a not uncommon occurrence—your carbide master will not grow at the same rate as your gage or parts. However, carbide is extremely corrosion resistant. Also, it has the highest wear resistance of any master material now in use. If your environment is so corrosive and violent that steel and even chrome plate do not hold up, carbide may be the answer.

What Kind Of Gage Do You Need?

A Baker's Dozen Factors To Consider

Like every other function in modern manufacturing operations, inspection is subject to management's efforts at cost control or cost containment. It's good business sense to try to maximize the value of every dollar spent, but it means that hard choices must be made when selecting gaging equipment. Issues as diverse as personnel, training, warranties, throughput requirements, manufacturing methods and materials, the end-use of the workpiece, and general company policies on gaging methods and suppliers may influence both the effectiveness and the cost of the inspection process.

For example, what's the ultimate cost of a bad part passing through the inspection process? It could be just a minor inconvenience to an OEM customer—maybe a two-second delay as an assembler tosses out a flawed two-cent fastener and selects another one. On the other hand, it could be a potentially disastrous equipment malfunction with expensive, even fatal, consequences. Even if the dimensional tolerance specifications for the parts are identical in both instances, management should certainly be willing to spend more on inspection in the second case to achieve a higher level of certainty—probably approaching 100 percent. One disaster averted will easily pay for the more expensive process in lawsuits avoided, lower insurance premiums, etc.

Many companies have achieved economies by moving inspection out of the lab and onto the shop floor. As this occurs, machinists and manufacturing engineers become more responsible for quality issues. Luckily, many gage suppliers are more than willing to spend time helping these newly assigned inspection managers analyze their functional requirements.

One could begin by comparing the hardware options. Let's take as an example a "simple" OD measurement on a small part. This inspection task could conceivably be performed with at least seven different gaging solutions:

1) Surface plate method, using V-blocks and test indicator
2) Micrometer
3) Purpose-built fixture gaging
4) Snap gage
5) Bench-type ID/OD gage with adjustable jaws

6) Hand-held air ring or air fork tooling
7) A fully automated system with parts handling.

(Actually there are many more solutions available, but let's keep it "simple.") Between these options there exists a price range from about $150 to $150,000. There are also differences in gage accuracy, operator influence, throughput, data output, and on and on. It's confusing, to say the least.

A better approach is to first define the functional requirements of the inspection task, and let that steer one toward the hardware that is capable of performing the tasks as identified. In order to do this, the end-user should consider the following factors:

- Nature of the feature to be inspected. Is it flat, round or otherwise? ID or OD? Is it easily accessible, or is it next to a shoulder, inside a bore, or a narrow groove?
- Accuracy. There should be a reasonable relationship between job tolerance and gage accuracy resolution and repeatability—very often on the order of a 10:1 ratio. A requirement for statistical GR&R (gage repeatability and reproducibility) testing may require 20:1. But always remember:
- Inspection costs. These increase sharply as gage accuracy improves. Before setting up a gaging operation for extremely close tolerance, verify that that particular level of accuracy is really necessary.
- Time and throughput. Fixed, purpose-built gaging may seem less economical than a more flexible, multi-purpose instrument, but, if it saves a thousand hours of labor over the course of a production run, it may pay for itself many times over.
- Ease of use, and training. Especially for shop floor gaging, you want to reduce the need for operator skill and the possibility of operator influence.
- Cost of maintenance. <u>Can</u> the gage be maintained, or is it a throwaway? How often is maintenance required, and who's going to perform it? Gages that can be reset to a master to compensate for wear are generally more economical over the long run than those that lose accuracy through extended use, but may require frequent mastering to ensure accuracy.
- Part cleanliness. Is the part dirty or clean at the stage of processing in which you want to measure it? That may affect labor requirements, maintenance, and the level of achievable accuracy, or it might steer you toward air gaging, which tends to be self-cleaning.
- Gaging environment. Will the gage be subject to vibration, dust, changes in temperature, etc.?
- "Mobility." Are you going to bring the gage to the part, or vice versa?
- Parts handling. What happens to the part after it's measured? Are bad parts discarded or reworked? Is there a sorting requirement?
- Workpiece material and finish. Is the part compressible? Is it easily scratched? Many standard gages can be modified to avoid such influences.
- Manufacturing process. Every machine tool imposes certain geometric and surface finish irregularities on workpieces. Do you need to measure them, or at least take them into consideration when performing a measurement?
- Budget. What do you have to work with?

All of these factors may be important when instituting an inspection program. Define as many as you can to help narrow the field, but remember that help is readily available from most manufacturers of gaging equipment—you just have to ask.

Gaging IDs And ODs

Without a doubt, circles are the most frequently produced machined form, generated by many different processes, including turning, milling, centerless grinding, boring, reaming, drilling, etc. There is, accordingly, a wide variety of gages to measure inside and outside diameters. Selecting the best gage for the job requires a consideration of many variables, including the size of the part, the length or depth of the round feature, and whether you want to gage in process or post process.

ID/OD indicator gages come in two basic flavors: benchtop and portable, as shown in Figures 1 and 2. Benchtop gages are generally restricted to measuring parts or features not more than 1" deep or long, while portable ID/OD gages can go as deep as 5" or so. If you need to measure hole IDs deeper than that, bore gages or plug gages are the tool of choice. On the other hand, snap gages are commonly used for ODs on longer parts—shafts, for example.

Getting back to ID/OD gages, the choice between benchtop and portable styles depends mainly on the size of the part being measured, and whether the part will be brought to the gage, or vice versa. If the part is large or awkward to manipulate, or if it's set up on a machine and you want to measure it there, then a portable, beam-type gage is required. Beam-type gages are available with maximum capacities from 5" to about 5', the largest ones

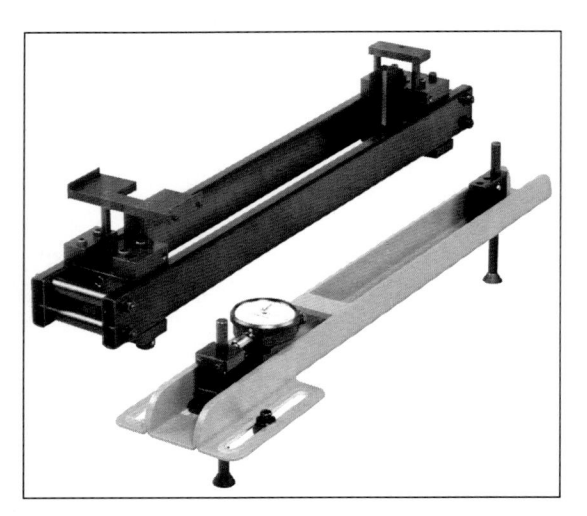

Fig. 1

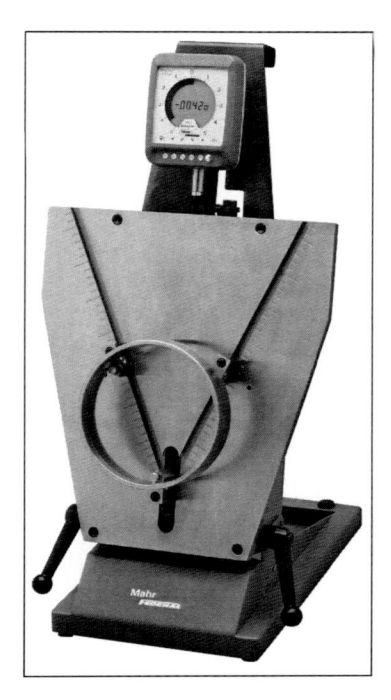

Fig. 2

being used to measure bearings and castings for jet engines and similarly large precision parts. Range of capacity is typically about 6", while the measurement range is determined by the indicator installed.

Most portable ID/OD gages lack centralizing stops, so they must be "rocked" like a bore gage to find the true diameter. When rocking the gage, use the fixed contact as the pivot, and allow the sensitive contact to sweep across the part. Likewise, if the gage must bear its own weight against the part, make sure that weight is borne by the fixed contact, not the sensitive one.

A special fixture with sliding stops at major increments is used to master for large ID measurements. Gage blocks are inserted in the fixture to "build out" the desired dimension. For OD measurements, calibrated "end rods" are often used: there is nothing especially fancy about these rods—they're simply lengths of steel, carefully calibrated for length. When mastering and measuring large dimensions, the gage, the master, and the part must all be at the same temperature. Otherwise, thermal influences will throw off the measurement.

Even so, don't expect very high precision when measuring dimensions of a foot or more. Most indicators on these large-capacity gages will have minimum grads of .0005". This is adequate, given the inability of most machine tools to hold tolerances much tighter than about .002" for parts that large. Beware the gage maker who tries to sell you a 3-foot capacity ID/OD gage with .0001" resolution: it's probably not capable of repeatable measurements.

Benchtop gages are used for smaller parts (diameters ranging from about .25" to about 9" maximum), and they're capable of higher precision. (.0001" is readily achievable.) There are two basic benchtop configurations: T-plates, and V-plates. A T-plate gage has sensitive and fixed contacts oriented normally, at 180° from each other, to measure true diameters. An extra fixed contact, oriented at 90° or 270°, serves to aid part staging. A V-plate gage has two fixed contacts offset symmetrically from the centerline, and the part is held against both of them. This arrangement requires a special-ratio indicator, because motion at the sensitive contact is actually measured relative to a chord between the fixed contacts, not to a true diameter.

This three-point arrangement is useful if the production process is likely to induce a three-lobed condition on the part—for example, if the part is machined in a three-jawed chuck. By rotating the part in a V-plate gage, one can obtain an accurate assessment of deviation from roundness. If the process is expected to generate an even number of lobes, then the T-plate layout is more appropriate to measure deviation.

Because they are self-centralizing, benchtop gages are capable of rapid throughput. To further accelerate gaging with either benchtop or portable gages, mechanical dial indicators can be replaced with electronic indicators. The dynamic measurement capabilities of the latest generation of digital indicators enable them to capture the minimum or maximum reading, or calculate the difference between those two figures. Operators are thus freed from having to carefully monitor the motion of a rapidly swinging needle on a dial indicator when rocking a portable gage, or checking for deviation on a benchtop version.

Gage Contacts: Get The Point?

In spite of their apparent simplicity, gage contacts represent a source of many potential measurement errors. When the simple act of touching a part can change its dimension, it's important to understand the ramifications of contact selection and application.

The first consideration must be whether you actually touch the part. Air gaging, as a non-contact method, has many advantages but is not always appropriate. Air gaging tends to average out surface finish variations on a part, providing a reading that lies between the microinch-height peaks and valleys. In some instances this may be desirable, but if the critical dimension lies at the maximum height on the surface, then contact gaging might be more appropriate.

Contact size and shape are critical. Contacts with small radii may nestle between high spots of surface and form irregularities, or might sit on top of them, depending on exactly where the gage contacts the workpiece. If the critical dimension is the low spot, it may be necessary to explore the part with the gage. Larger radii or flat contacts will bridge across high spots. The choice of radius depends at least partly on whether you want to "ignore" surface and geometry irregularities on the high or low side.

Contact size and shape also influence measurements because all materials compress to some extent as a function of pressure. When measuring obviously compressible materials such as plastics or textiles, gaging practice is commonly guided by industry standards. For example, ASTM D-461, "Standard Methods of Testing Felt," specifies the size of the bottom anvil (min. 2 in^2), the size and shape of the upper contact ($1 = 0.0001$ in^2; i.e., 1.129" diameter, with edge radius of $0.016 = 0.001$ in^2), the force of the contact ($10 = 0.5$ oz.), and the amount of time allowed for material compression prior to taking the measurement (minimum 10 secs.). Similarly detailed standards exist for measuring the thickness of wire insulation, rubber sheet stock, and dozens of other materials. Not all the contacts defined in the standards are flat, parallel surfaces: other shapes such as knife-edges, buttons, cylinders, or spheres may be specified.

Even materials that are not thought of as compressible do compress somewhat under normal gaging pressures. Because of the higher and higher levels of accuracy required in metalworking industries, it is often essential to compensate for this.

Under a typical gaging force of 6.4 ounces, a diamond-tipped contact point with a radius of 0.125" will penetrate a steel workpiece or gageblock by 10 microinches. The same contact will compress tungsten carbide by 6.6 microinches, and fused quartz by 20 microinches. If microinches count in your application, it is important that workpiece and master be of the same material. Alternately, one can refer to a compensation table to make the necessary adjustment to the gage reading. Compression can be minimized by using a contact with larger surface area.

Contact material also makes a difference. For the sake of durability, one normally selects a contact point that is harder than the workpiece. Typical choices include (in increasing order of hardness): hardened steel, tungsten carbide, and jeweled tips—ruby, sapphire, or diamond. Tungsten carbide is a good choice for measuring steel parts unless millions of cycles are anticipated, in which case diamond might be chosen for longer life. One should avoid using tungsten carbide contacts on aluminum parts, however. Aluminum tends to

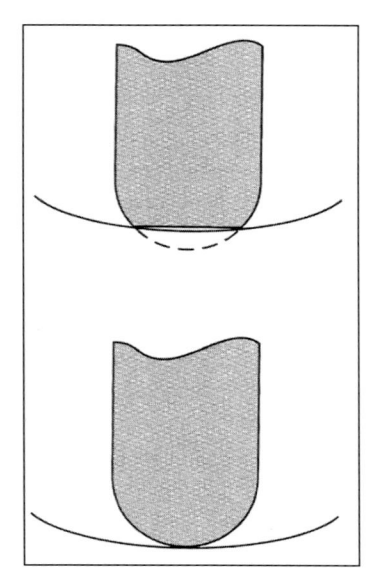

Fig. 1

"stick" to carbide, and it can build up so quickly as to throw off measurements between typical mastering intervals. Hardened steel or diamond is a better choice for measuring aluminum.

As shown in Figure 1, differently shaped parts may produce different readings, even though they are dimensionally identical. This is especially true when the contact points are worn. It is often possible to obtain accurate gage readings with worn contacts if one masters carefully and frequently. This includes using a master that is the same shape as the workpiece. Periodically confirm that the gage contacts are parallel by sliding a precision steel ball or wire on the anvil from 12 o'clock to 6 o'clock, and from 3 o'clock to 9 o'clock, and measuring for repeatability. Measure again with the ball in the middle of the anvil to check for wear there.

Make sure the contact is screwed firmly into its socket so there is no play. On rare occasions, a jeweled insert may come slightly loose in its steel holder. A simple repeatability check will detect this. Unfortunately, there's no good fix for it. Give the diamond to your sweetheart, and install a fresh contact.

Not all gages use perpendicular motion. If yours has angular motion, be aware that

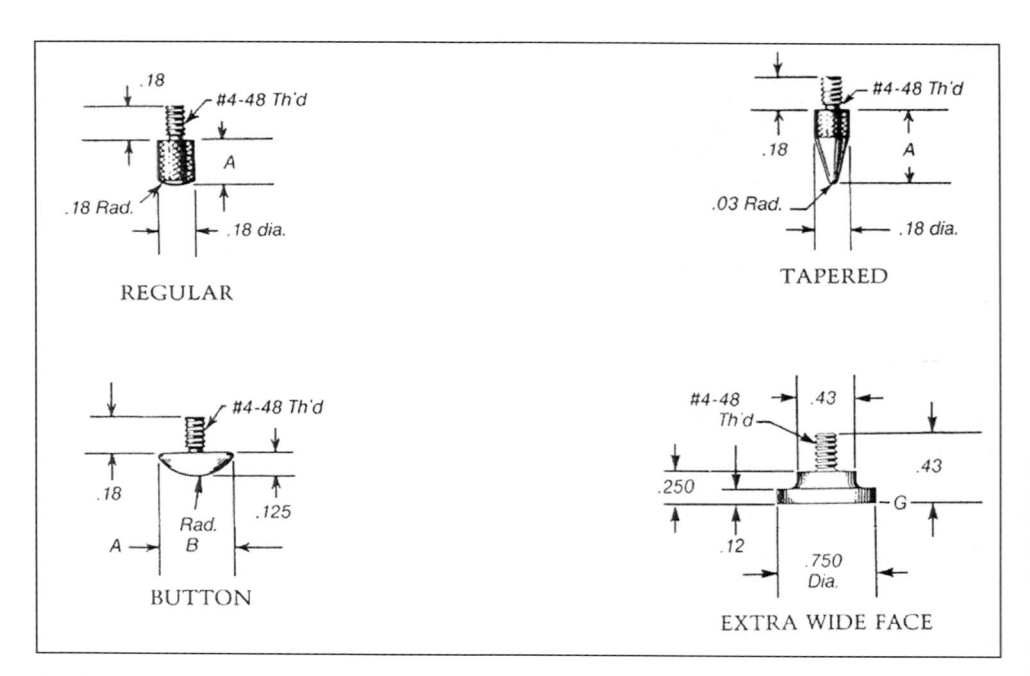

REGULAR

TAPERED

BUTTON

EXTRA WIDE FACE

Fig. 2

34   Quality Gaging Tips

changing the length of the lever contact will change the reading. On mechanical test indicators, you may be able to install a new dial face with the proper magnification, or you can apply a simple mathematical compensation to every measurement. If you're using a lever-type electronic gage head, you might be able to program the compensation into the amplifier.

A Physical Check-Up For Gages

Just like the people who use them, gages should have periodic physical examinations. Sometimes, gage calibration is needed to identify the seriousness of a known problem, and sometimes it uncovers problems you didn't know existed. But as with a people-exam, the main reason for the annual check-up is to prevent problems from occurring in the first place.

The accuracy of a gage can only be known by reference to a higher standard. Thus, gages are set to masters that are more accurate than the gage. These masters are certified against gage blocks produced to a higher standard of accuracy—ultimately traceable to nationally or internationally recognized "absolute" standards that define the size of the dimensional unit. This is the line of traceability, which must be followed for calibration to be valid.

Calibration is used to determine how closely a gage adheres to the standard. When applied to a master ring, disc, or a gage block, it reveals the difference between the nominal and the actual sizes. When applied to a measuring instrument such as a comparator, calibration reveals the relationship between gage input and output—in other words, the difference between the actual size of the part and what the gage says it is.

Gages go out of calibration through normal usage: parts wear, and mechanisms become contaminated. A gage may have a design flaw, so that joints loosen frequently and the gage becomes incapable of holding calibration. Accidents and rough handling also put gages out of calibration.

No gage, therefore, can be relied upon if it has not been calibrated, or if its calibration history is unknown. Annual calibration is considered the minimum, but for gages that are used in demanding environments, gages that are used by several operators or for many different parts, and gages used in high-volume applications, shorter intervals are needed. Frequent calibration is also required when gaging parts that will be used in critical applications, and where the cost of being wrong is high.

Large companies that own hundreds or thousands of gages sometimes have their own calibration departments, but this is rarely an economical option for machine shops. In addition to specialized equipment, in-house calibration programs require a willingness to devote substantial employee resources to the task.

Calibration service providers are usually a more economical approach. Smaller gages can be shipped to the provider, while large instruments must be checked in-place. Calibration houses also help shops by maintaining a comprehensive calibration program, to ensure that every gage in the facility is checked according to schedule, and that proper records are kept.

General guidelines to instrument calibration procedures appear in the ISO 10012-1 and ANSI Z540-1 standards. While every gage has its own specific procedures, which are outlined in the owner's manual, calibration procedures also must be application-specific. In other words, identical gages that are used in different ways may require different procedures.

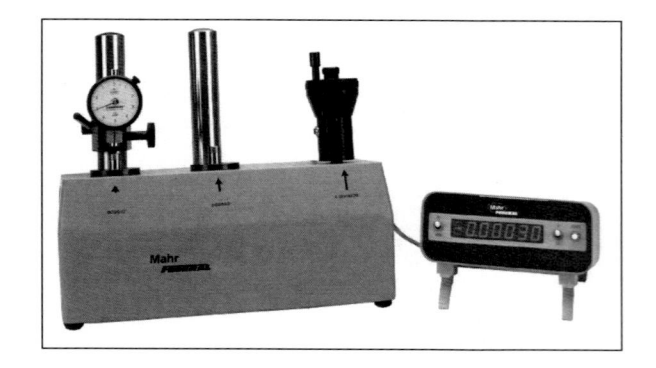

For example, if a gage is used only to confirm that parts fall within a tolerance band, it may be sufficient to calibrate it only at the upper and lower tolerance limits. On the other hand, if the same gage is used to collect data for SPC, and the accuracy of all measurements is important, then simple calibration might be insufficient, and a test of linearity over the entire range might be needed.

The conditions under which calibration occurs should duplicate the conditions under which the gage is used. Calibration in a high-tech gaging lab may be misleading if the gage normally lives next to a blast furnace. Similarly, a snap gage that is normally used to measure round parts should be calibrated against a master disc or ring, and not with a gage block. The gage block could produce misleading results by bridging across worn areas on gage contacts, while a round master would duplicate the actual gaging conditions and produce reliable results.

Before calibration begins, therefore, the technician should be provided with a part print and a description of the gaging procedure. Next, he should check the calibration record, to confirm that the instrument serial number and specifications agree with the instrument at hand. The gage will then be cleaned and visually inspected for burrs, nicks, and scratches. Defects must be stoned out, and mechanisms checked for freedom of movement. If the instrument has been moved from another area, it must be given time to stabilize.

All of these measures help ensure that calibration will be accurate, but this must not lead to a false sense of security: gage calibration will not eliminate all measuring errors. As we have seen before, gaging is not simply hardware: it is a process. Calibration lends control over the instrument and the standard or master, but gage users must continue to seek control over the environment, the workpiece, and the gage operator.

■ ■ ■

Squeezing More Accuracy From A Gaging Situation

All gages are engineered to provide a specified level of accuracy under certain conditions. Before specifying a gage, users must take stock of all the parameters of the inspection process.

How quickly must inspection be performed? Many gages, which are capable of high levels of accuracy, require careful operation to generate reliable results. Others are more foolproof, and can generate good results more quickly, and with less reliance on operator skill.

Where will inspection take place? Some gages are relatively forgiving of environmental variables—for example, dust, cutting fluid residues, vibration, or changes in temperature—while others are less so. Likewise with many other factors in the gaging situation. The ability to obtain specified accuracy from a gage in a real inspection situation depends upon the prior satisfaction of many parameters, both explicit and assumed.

Recently, a manufacturer came to me with a requirement to inspect a wide variety of hole sizes on a line of 4-liter automotive engines. Some of the relevant parameters of the gaging situation included:

- **Throughput**. With literally hundreds of thousands of parts to measure, inspection had to be fast and foolproof.
- **Output**. The manufacturer required the capability of automatically collecting data for SPC.
- **Portability**. The parts being gaged were large, so the gage had to come to the parts, not vice versa.
- **Accuracy**. Most hole tolerances were ±0.001", but some were as tight as ±0.0005".

Adjustable bore gaging wouldn't do the job, because of slow operation and a high requirement for operator skill. Air gaging, while fast and sufficiently foolproof, was not sufficiently portable for the application. We settled on fixed-size mechanical plug gaging, equipped with digital electronic indicators to provide data output.

The manufacturer specified a GR&R (gage repeatability and reproducibility) requirement of 20% or better on holes with tolerances of ±0.001". This meant that the system had to perform to 80 microinches or better. This requirement was met using standard gage plugs, and standard digital indicators with resolution of 50 microinches: GR&R achieved with these setups was ≤16%.

On holes with tolerances of ±0.0007" and ±0.0005", however, the manufacturer required GR&R of 10%, which translated to 40 microinches. Given the other parameters of the application, mechanical plug gages remained the only practical approach, so we had to find a way to "squeeze" more accuracy out of the situation.

Plug gages are typically engineered for 0.002" of material clearance in the holes they are designed to measure, to accommodate undersize holes, and to ease insertion. The greater the clearance, the greater the amount of centralizing error, in which the gage measures a chord of the circle, and not its true diameter. By reducing the designed clearance, centralizing error can be minimized—with some tradeoff against ease of insertion.

We engineered a special set of plug gages, with minimum material clearance of 0.0007". The standard digital indicators were also replaced with high-resolution units, capable of 20 microinch resolution. This combination satisfied the requirements, generating GR&R of ≤8.5%.

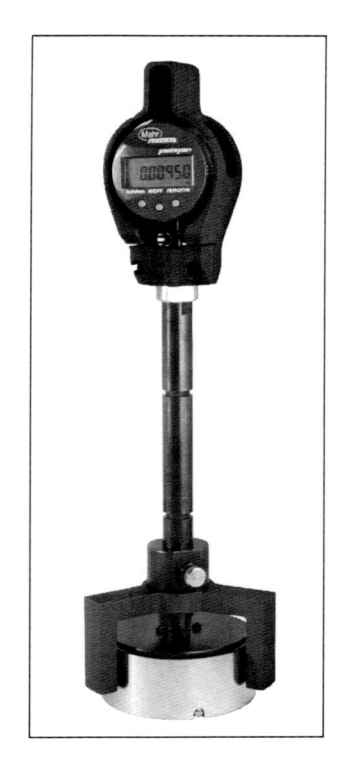

Carefully considered changes in both gage design and gaging practice can help to squeeze the greatest amount of accuracy from a standard gage.

Section B: Basics of Measurement

Remember SWIPE? This acronym stands for the five categories of gaging variables: Standard (i.e., the master); Workpiece; Instrument (i.e., the gage); Personnel; and Environment. In the case of the engine manufacturer, we tweaked the instrument, thus reducing one source of gaging variability. We reduced a second source by providing higher-quality masters for these gages. If throughput had not been such a high priority, we might have considered altering the environment where inspection was performed, or providing more training to personnel. If portability hadn't been an issue, then the solution might have been a different instrument altogether.

The five categories of gaging variables encompass dozens of specific factors. (For example, within the category of Workpiece, there are variables of surface finish and part geometry that may influence dimensional readings.) To squeeze more accuracy out of a gaging situation, look for opportunities to reduce or eliminate one of more of these factors.

The Real Dirt About Gaging

I am not sure that any of us in the metrology business are very close to godliness, but I do know that cleanliness is the first step to approaching accuracy in gaging. Probably every machinist is at least nominally aware that dirt can interfere with the ability to take accurate measurements. But the importance on the issue cannot be over-emphasized, and even a conscientious user can occasionally use a reminder.

Leave your gage out of its box for a few hours. Then check it for zero setting. Next, clean the measuring surfaces and blow off the lint. Check the zero setting again. You will probably find a difference of about 0.0005" due to dirt on these surfaces.

Test number two: We left a clean master disc, marked 0.7985"XX unprotected for a number of hours on a work bench in the shop. Then, taking special pains not to touch its measuring surfaces, we brought it into a temperature-controlled room and let it cool off before measuring it with an electronic comparator. The needle went off the scale, which meant that the master plus dirt was more than 0.0003" larger than the nominal 0.7985" setting. Then we carefully and thoroughly cleaned the master with solvent and measured it again. The reading was +0.000004 from nominal.

Finally, we cleaned the master again, using the time-honored machinist's method of wiping it with the palm of the hand. Measuring again, it had gone up to +0.000013". We lost half the normal gage tolerance by "cleaning" it with the palm. (Some slight error may also have been introduced through expansion of the master due to conductive heating from the hand. (More on this subject in a later paragraph).

We have already seen how dirt in invisible quantities can skew a measurement, both on the contacting surfaces of the gage itself and on the workpiece or master. And recall that our examples were reasonably clean environments. Now picture the common abode of a gage in your shop: Is it living in an oil-and-chip-filled apron of a lathe or screw machine, or perhaps sharing the pocket of a shop apron with pencil stubs, pocket lint and what have you?

Aside from simply getting in the way of a measurement, dirt also impedes accurate measurement by increasing friction in a gage's movement. Drag may prevent a mechanism from returning to zero, and every place that friction must be overcome represents a potential for deflection in the gage or the setup. If dirt is the biggest enemy of accurate measurement, then friction is a close second.

Next time you have a serviceman in to work on a gage, watch him. Chances are, the first thing he does is clean the gage, whether it is a simple dial indicator or a Coordinate Measuring Machine. If you take only one thing away from this column, this should be it: eliminate dirt as a possible source of error before attempting to diagnose a malfunctioning gage.

■ ■ ■

Gage Layout Is Up To The User

Previous columns have discussed how most dimensional gaging applications are really just variations on four basic themes, to measure height, depth, thickness, or diameter. Relational gaging applications are nearly as straightforward, conceptually. Measuring qualities like roundness, straightness, squareness, taper, parallelism, or distance between centers is usually a matter of measuring a few dimensional features, then doing some simple calculations.

Better yet, let the gage do the calculations. Even simple benchtop gaging amplifiers can measure two or more dimensions simultaneously and manipulate the readings through addition, subtraction, or averaging. (Air gaging can also be used in many of these applications, but for simplicity, we'll stick with electronic gage heads as the basis of discussion.) As shown in the schematics of Figure 1, a wide range of relational characteristics can be measured with just one or two gage heads: it's basically a question of setting them up in the right configuration—and mak-

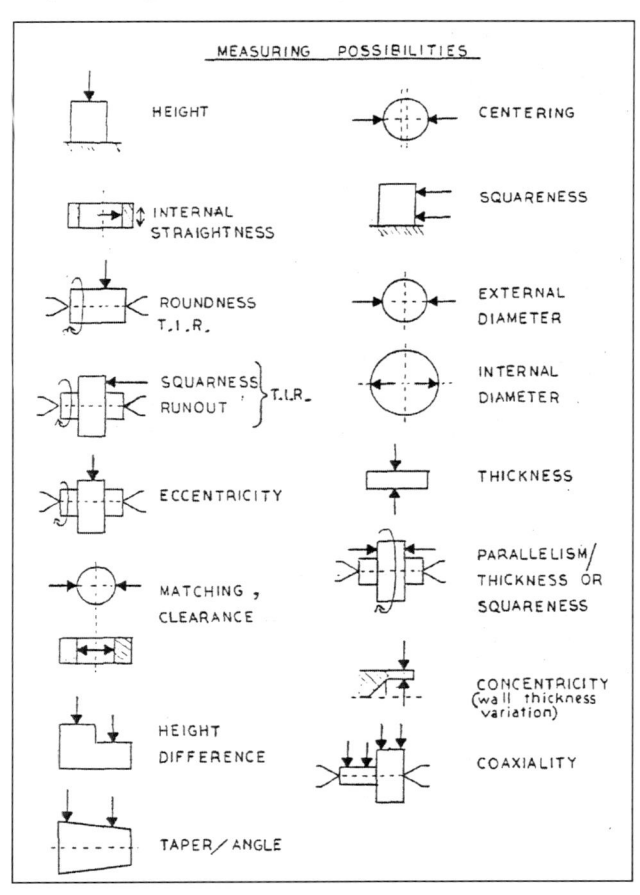

Fig. 1

Section B: Basics of Measurement   39

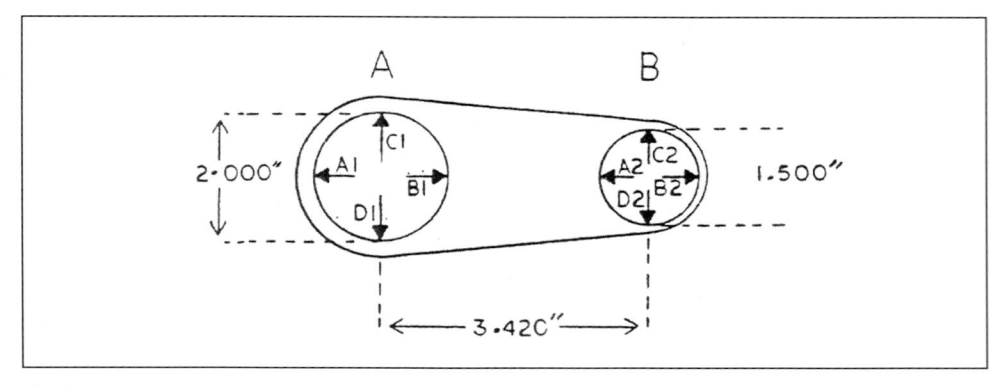

Fig. 2

ing sure that the fixture is capable of maintaining a precise relationship between the part and the gage heads.

With a little imagination, you can combine several related and/or independent measurements into a single fixture to speed up the gaging process. Figure 2 shows a fixture gage for measuring connecting rods. Transducers A1 and B1 measure the diameter of the crank bore: out-of-roundness can be checked by comparing that measurement with a second diameter at 90° (C1 and D1). The same features are measured on the pin bore, using transducers A2 through D2. Finally, the distance between bore centers can be calculated, using the same gaging data.

Using these principles, machine shops can develop workable fixture gages in-house for a wide range of applications, or modify existing gages to add capabilities. Electronic gage heads (i.e., transducers) and air probes are available in many configurations and sizes, some of them small enough to permit simultaneous measurements of very closely-spaced part features. Before you begin in earnest, you'll need to check the manufacturer's specs for gage head dimensions, accuracy, and range. Even if you don't want to build the gage in-house, you can use these ideas to design a "schematic" gage to aid you in your discussions with custom gage makers.

■ ■ ■

Stage It To Gage It

Freedom is not always a good thing, at least when it comes to gaging. Some gaging applications call for inspecting a part for variation across a given feature, which calls for freedom of movement in at least one plane. Other applications call for measuring a series of parts at exactly the same location on the feature, time after time. In the first instance, you're checking the accuracy of the part. In the second, where you're checking the repeatability of the process, freedom of movement is the enemy.

For example, to inspect a nominally straight bore for taper error, using an air gaging plug or a Dimentron®-type mechanical plug, insert the plug slowly, and watch the indicator needle or readout display for variation as you go. On the other hand, if you are inspecting

IDs to confirm the stability of the boring process from part to part, you must measure every bore at exactly the same height. If you do not, any taper present may lead you to an erroneous conclusion that the process is unstable. The first application requires freedom of axial movement. The second requires that axial movement be eliminated. This can be readily done by installing a stop collar on the plug to establish a depth datum.

The number and type of datums required varies with the type of gaging and the application. Figure 1 shows a fixture gage to check a piston for perpendicularity of the wrist pin bore to the piston OD. (Piston skirts are typically ovoid: this is shown exaggerated. The skirt's maximum OD equals the head OD, which is round to the centerline of the pin bore.) The bore is placed over an air plug, which serves as a datum, locating the part both lengthwise and radially. The critical element in the engineering of the gage is in the dimensioning of the two V-blocks that establish the heights of both ends of the part. Because of the skirt's ovality, the V-block at that end must be slightly higher, to bring the OD of the head of the piston perpendicular to the plug. Without reliable staging in this plane, the gage could not generate repeatable results.

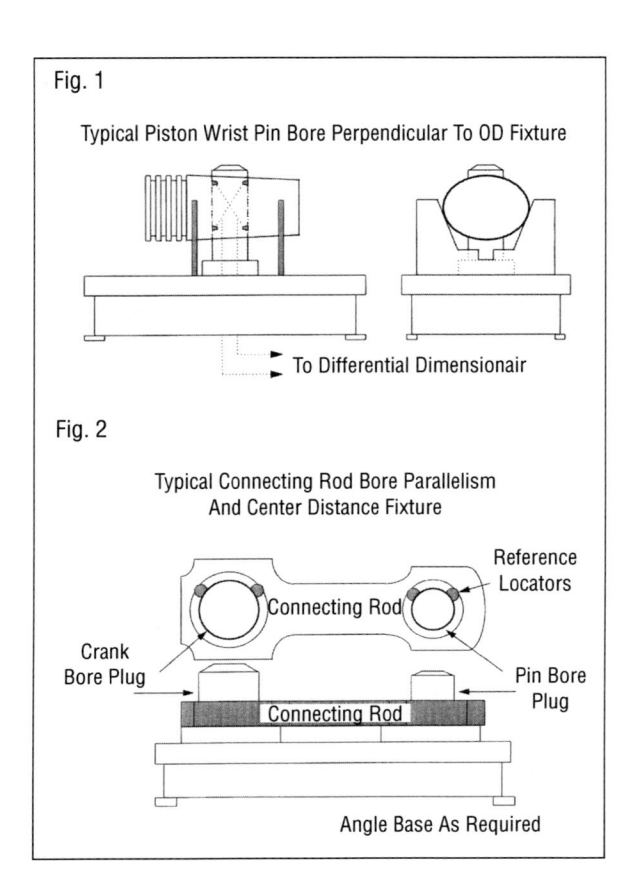

As many as three datums may be required to properly locate a part and a gage relative to one another in three dimensional space. Refer to Figure 2. This air fixture gage checks connecting rod crank and pin bores for parallelism (bend and twist) and center distance. Placing the conrod flat on the base establishes the primary datum. Although it is not shown in the diagram, the base is angled several degrees toward the viewer: the uppermost ODs of the plugs therefore establish a secondary datum against which the conrod rests. Two precision balls are installed on each plug, located at a height half the depth of the bores. These balls locate the part lengthwise, establishing a tertiary datum.

Before a fixture gage can be designed, the engineer must understand what specifications are to be inspected. In many respects, the design of the gage mirrors not only the design of the part, but also the manufacturing processes that produced it. Machinists must establish datums in order to machine a part accurately, and gage designers often need to know what those datums were, in order to position the part repeatably relative to the gage head or other sensitive device. When working with a custom gage house, therefore, operation sheets should be provided, in addition to part prints. If you're working with an in-house "gage maker" or a less experienced supplier, make sure that the staging is designed around a careful analysis of the part and processes.

Fixtures Are A Common Source Of Gaging Error

As a gaging engineer, my concept of a gage includes both the measuring instrument and its fixture. Assuming you are dealing with a reputable supplier and your instrument was engineered to do its job as intended, there is probably little you can do to improve its accuracy, aside from throwing it out and spending more money. So we will concentrate on the setup, which is a common source of measurement errors.

The fixture establishes the basic relationship between the measuring instrument (that is, a dial indicator) and the workpiece, so any error in the fixture inevitably shows up in the measurements. Many fixtures are designed as a variation of a C-frame shape and, as such, have a substantial cantilever that is subject to deflection. This problem is greatly reduced if the fixture is a solid, one-piece unit.

Most fixtures, however, consist of a minimum of three pieces: a base, a post, and an arm. These components must be fastened together with absolutely no play between them. As a rough rule of thumb, any movement between two components will be magnified at least tenfold at the workpiece. Play of only a few millionths can, therefore, easily accumulate through a couple of joints so that measurements to ten-thousandths become unreliable, regardless of the level of discrimination of the instrument.

Because such tight tolerances are required—tighter than you can perceive by eye or by touch—it is often essential that fixtures have two setscrews per joint. No matter how tightly a single setscrew is tightened, it often acts merely as a point around which components pivot.

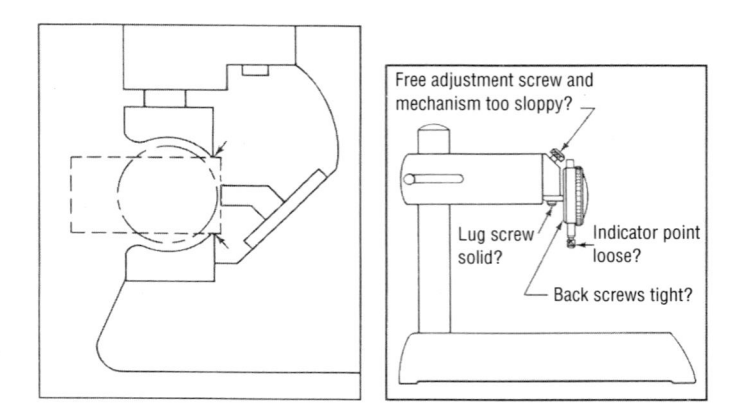

Left—After repeated measurements, round workpieces may create wear in the measurement surface. Size setting with the gage block may detect this wear, bridging these surfaces.
Right—Loose screws are a common source of gaging error.

· Lost motion due to play between fixture components is dangerous. Assuming that the gage is mastered regularly, a fixture with loose joints may still provide accurate comparative measurements. There are two places in a gage, however, where loose assembly may produce erratic readings, making the setup completely unreliable. Most dial indicators offer optional backs and sensitive contacts that are designed to be changed by the end-user. Looseness of these two components is among the most common sources of gaging error. These are often the first places a gage repairperson looks to solve erratic readings.

Fixtures must be designed to position workpieces consistently in relation to the measuring instrument. This is critical if the master is a different shape from the workpiece. For instance, when using a flat gage block to master an indicator that is used to check ODs on round workpieces, the fixture must position the workpiece to measure its true diameter—not a chord.

The use of masters that are the same shape as the workpiece avoids this problem and another one that can be more difficult to isolate. After repeated measurements, round workpieces may wear a hollow, allowing accurate comparative measurements, while flat gage blocks may bridge the wear, introducing a source of error.

Regardless of its complexity, your gage fixture is the key to accurate measurements. Make sure there is no play at its joints. Check that the instrument, itself, is assembled securely. And confirm that the gage measures workpieces and masters at identical locations.

Gaging Accuracy Is Spelled S-W-I-P-E

When a gaging system is not performing as expected, we often hear the same dialogue. The operator, who has only his gage to go by, says, "Don't tell me the parts are no good—they measure on my gage." The inspector replies, "Well, the parts don't fit, so if your gage says they are okay, your gage is wrong."

This is the natural reaction. People are quick to blame the instrument because it is easy to quantify. We can grab it, take it to the lab and test it. However, this approach will often fail to find the problem, or find only part of it, because the instrument is only one-fifth of the total measuring system.

The five elements of a measuring system can be listed in an acronym, SWIPE, and rather than immediately blaming the instrument when there is a problem, a better approach is to examine all five elements:

S represents the standard used when the system is set up or checked for error, such as the master in comparative gages of the leadscrew in a micrometer. Remember, master disks and rings should be handled as carefully as gage blocks, because nicks and scratches can be a significant contributor to error.

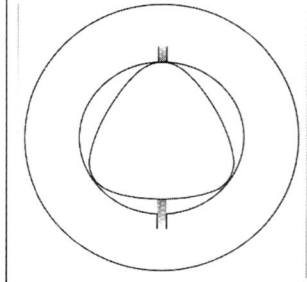

Variations in geometry and surface finish of the workpiece can directly affect a size measurement. For example, when measuring a centerless-ground part with a two-jet air ring, a three-point out-of-round condition will not show up because you are only seeing average size.

Section B: Basics of Measurement   43

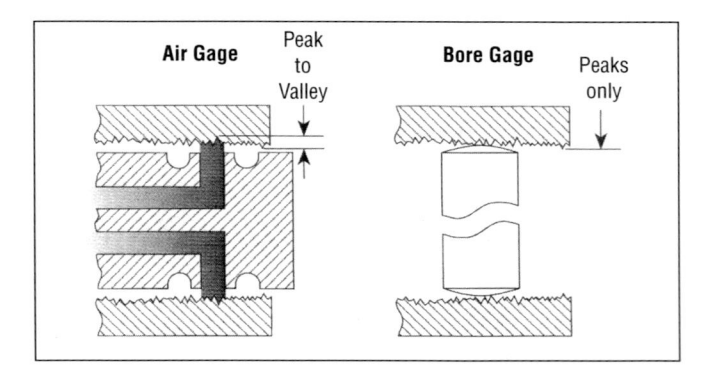

Air Gage — Peak to Valley

Bore Gage — Peaks only

Remember to use the same gage as your customers, or you could end up blaming each other's instruments. For example, checking bores with an air gage gives the average of the peaks and valleys, while the customer's mechanical gage says the bores are too small because it only sees the peaks.

W is the workpiece being measured. Variations in geometry and surface finish of the measured part directly affect a system's repeatability. These part variations are difficult to detect, yet can sometimes manifest themselves as apparent error in the measuring system. For example, when measuring a centerless ground part with a two-jet air ring, a three-point out-of-round condition will not show up because you are only seeing average size.

I stands for the instrument itself. Select a gage based on the tolerance of the parts to be measured, the type of environment and the skill level of the operators. And remember what your customers will be measuring the parts with. Say, for example, you are checking bores with an air gage, but your customer inspects them with a mechanical gage. Since the surface is not a mirror finish, your air gage is giving you the average of the peaks and valleys, while the customer's mechanical gage is saying the bores are too small because it only sees the peaks. Neither measurement is "wrong", but you could end up blaming each other's instruments.

P is for people. Failure to adequately train operating personnel will ensure poor performance. Even the operation of the simplest of gages, such as air gaging, requires some operator training for adequate results. Most important, the machine operator must assume responsibility for maintaining the instruments. Checking for looseness, parallelism, nicks and scratches, dirt, rust, and so on, is absolutely necessary to ensure system performance. We all know it, but we forget when we are in a hurry.

E represents the environment. As I have said before in this column, thermal factors such as radiant energy, conductive heating, drafts and room temperature differentials can significantly impact gage system performance. And, again, dirt is the number one enemy of gaging. So the problem that has you pulling your hair out and cursing your instruments could be as simple as your shop being a little warmer or a little dustier than your customer's.

Before blaming your gage, take a SWIPE at it and consider all the factors influencing its accuracy.

■ ■ ■

Central Intelligence

Many factors influence the accuracy of hole diameter measurements. We've seen the importance of operator skill in the use of rocking-type adjustable bore gages, and discussed how variations in part geometry may make even technically accurate measurements inaccurate from a part-function perspective.

One of the fundamental requirements in bore gaging is that the gage contacts be centered in the bore. Bore gages that are not properly centered measure a chord of the circle, rather than its true diameter. Operator error is a common cause of poor centralization with rocking-type gages, while wear or damage can affect the centralization of any gage.

Most adjustable bore gages have a centralizer that helps the operator align the gage properly. Through misuse or wear, a centralizer may be damaged, so that the reference contact is pushed off-center. As long as the centralizer is not loose, it may still be possible to master the damaged gage with a ring gage: the off-center relationship will probably carry over to workpieces, so repeatable results might be obtained. Errors in part geometry, however, could cause a lack of agreement between results from the damaged gage and an undamaged one. And if the damaged gage were to be mastered with gage blocks on a set-master, a different zero reading would be obtained. So in spite of the possibility that an adjustable bore gage with a damaged centralizer might generate accurate results, it cannot be relied upon.

Fixed-size bore gages, such as air tooling and mechanical plugs, are substantially self-centering. They are engineered with a specified clearance between the gage body and a nominal-size bore that is a compromise between ease of insertion on the one hand and optimum centralization on the other. But after years of use, the plug may become worn, resulting in excessive clearance and poor centralization.

Checking centralization is easy for both gage types. For rocking-type gages, simply compare measurements between a master ring and a set-master of the same nominal dimension. The difference between the round and square surfaces will reveal any lack of centralization. To check a two-contact or two-jet plug, insert the gage horizontally into a master ring, allowing the master to bear the gage's weight. Measure once with the contacts or jets oriented vertically, and once horizontally. If the measurements differ, centralization is poor.

<u>Centralization error</u> is the difference between the true diameter and the length of the chord measured. Quality personnel occasionally specify centralization error as a percentage of the total error budget (or repeatability requirement) for a gaging operation. For example, the error budget might be 10 percent of the tolerance: in addition to an allowance for centralization error, this might include influences of operator error; gage repeatability; environmental variation; and within-part variation (e.g., geometry error).

Gage users should be prepared to calculate how far off the bore centerline a gage may be without exceeding the specified centralization error. We'll call the allowable distance between the bore centerline and the contact centerline the <u>misalignment tolerance</u>. A simple formula, based on the relationship between the legs and the hypotenuse of a right triangle, does the job:

$$x^2 = z^2 - y^2$$

where:
x = misalignment tolerance
$y = z - 1/2$ centralization error
$z = 1/2$ nominal diameter

```
1/2 centralization error
         |
         |
       y
         |

                    — x —

x = misalignment tolerance
y = z - 1/2 centralization error
z = 1/2 nominal diameter

                                — z

                          ℄ Contacts
                    ℄ Bore
```

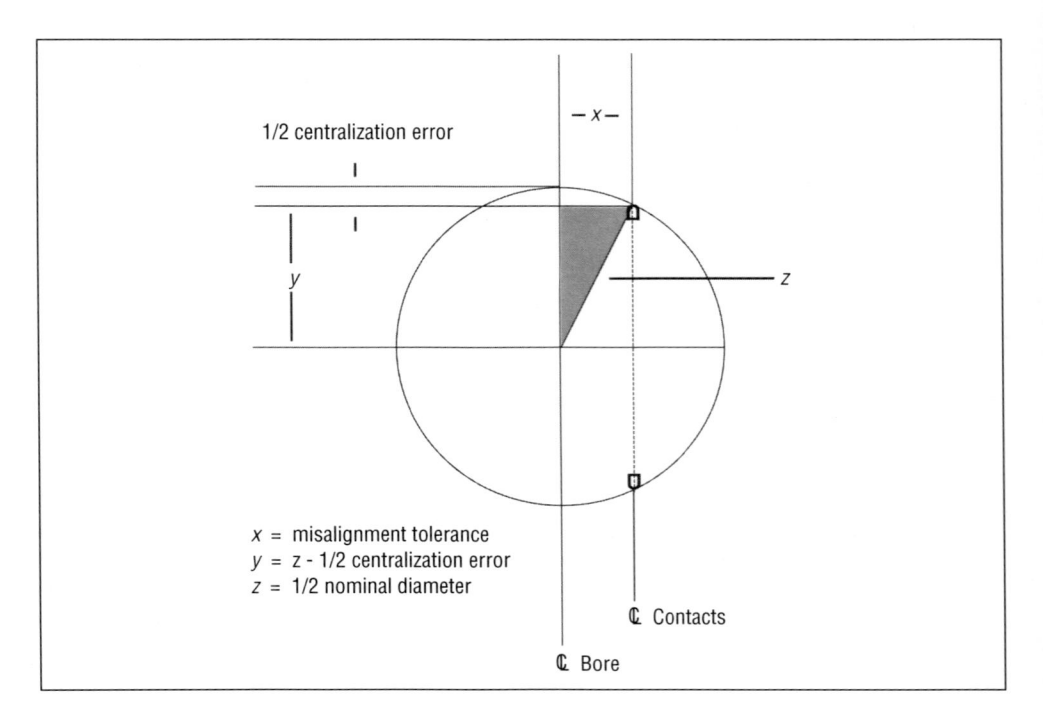

Let's run through an example. The nominal bore dimension is .5", with a dimensional tolerance of .002" (±.001"). Centralization error is specified at a maximum of 2 percent of the dimensional tolerance (or .02 × .002" = .00004").

$z = .5" \div 2 = .250"$
$y = .250" - (.00004" \div 2) = .24998"$
$x^2 = (.250")^2 - (.24998")^2$
$x^2 = .0625" - .06249"$
$x^2 = .00001"$
$x = .00316"$

The gage can be misaligned slightly more than .003" off-center before it will exceed the allowable centralization error. If you run through the same exercise for a 5.0" nominal bore, keeping the other values constant, you'll find that misalignment can be up to .010" before centralization error exceeds 2 percent of the .002" dimensional tolerance. Thus, as bore size increases, so does the misalignment tolerance.

Take A Stand

Bench comparators consist of an indicating device, plus a height stand that incorporates a locating surface for the part. (In contrast, a height stand that has no locating surface is known as a test stand, and must be used with a surface plate.) There are hundreds of bench comparator stands on the market, so it's important to understand their features and options.

On some stands—especially those used to measure large parts—the base itself serves as the reference surface. Bases may be either steel or granite, with steel being easier to lap flat when necessary.

For higher accuracy, it is usually desirable to use a comparator stand with a steel or ceramic anvil mounted to the base. As a smaller component, the anvil can be machined to a tighter flatness tolerance than the base—often to a few microinches over its entire surface. In some cases, the anvil may be so flat as to provide a wringing surface for the workpiece—an excellent condition for very critical measurements. The anvil is also easier to keep clean, and can be more readily adjusted to squareness with the indicator.

Some anvils have diagonal grooves milled into the reference surface. These serrations serve to wipe any dirt or chips off the part, and reduce the contact area across which contamination-induced errors may occur.

Accessory positioning devices may be used to increase the comparator's repeatability in various applications. A flat backstop permits lateral exploration of the part for variation, while a vertical vee used as a backstop permits rotational exploration of round parts. A vee can also be mounted horizontally, thus serving as a reference in two directions. Round workpieces may also be held horizontally between a pair of centers attached to the base for runout inspection. A round, horizontal arm may be attached to the post, below the arm that holds the indicator, to serve as a reference for measuring the wall thickness of tubes. And special fixtures may be designed to position odd-shaped parts without a flat bottom. The post is the next important component, where bigger and heavier usually means more stability and less variability. Some posts have spiral grooves to reduce the chance of dirt getting between the post and the arm clamp, which is an invitation to part wear and "slop" in the setup.

The post should provide some kind of arm support beyond the arm's own clamp. Without it, you risk dropping the arm every time you loosen the clamp to adjust the height, which could result in damaged components and crunched fingers. At minimum, there should be a locking ring on the post. A better approach is a rack and pinion drive, which makes it much easier to position the arm, especially if it's a heavy one. Even these should be equipped with their own locking mechanism, so that the weight of the arm does not constantly rest on the drive screw. In some cases, the "post" may be a dovetail slide, which eliminates rotation of the arm in the horizontal plane. This can make setup easier when the anvil remains the same, but the arm must be raised or lowered to measure parts of different lengths.

When it comes to the arm, shorter is better to minimize flexing, although a longer arm may be needed for larger parts. A fine height adjustment screw is a valuable feature for accurate positioning of the indicator relative to the part. Also look for a good locking device that clamps the post to the arm across a broad surface rather than at a single point, as this could allow the arm to pivot up and down. An axial swiveling feature is available with some arms for special positioning needs.

As simple as comparator stands may be, there are hundreds of options, sizes, and levels of quality. Take the time to understand your application thoroughly, and make sure you buy enough capabilities for your needs. You'll end up with faster, easier, more accurate measurements, and less time spent on repairs and adjustments. It may cost more initially, but you'll come out ahead.

Perfect Gaging In An Imperfect World

It is certainly not news that, more and more, gages are being forced out onto the shop floor. Tight-tolerance measurements that were once performed in a semi-clean room by a trained inspection technician are now being done right next to the machine, often by the machinist. But just because shop floor gaging has become commonplace, doesn't mean that just any gage can be taken onto the shop floor. To assure good gage performance, there are a number of specifications and care issues, which need to be addressed.

Is the gage designed to help the user get good measurements? A gage with good Gage Repeatability and Reproducibility (GR&R) numbers will generate repeatable measurements for anyone who's trained to use it properly. Technique or "feel" should have minimal impact on results.

Gages with good GR&R are typically very robust. Part alignment is designed in, to make sure the part is held the same way every time and eliminate the effects of operator influence on part positioning. Bench gages usually outperform handheld gages in both respects.

Is the gage designed for the rigors of the shop floor environment? Gages designed for laboratory use often cannot cope with the dust and oil present on the shop floor. Some features commonly found on good shop floor gages include: careful sealing or shielding against contaminants; smooth surfaces without nooks and crannies that are difficult to clean; and sloping surfaces or overhangs designed to direct dust and fluids away from the display. (Try to distinguish between cabinets that just look good, and those that are truly functional.) If all of these are absent, one can often add years to a gage's useful life by installing it in a simple Lexan enclosure with a hinged door, or even by protecting it with a simple vinyl cover when it's not in use.

Is the gage easy to operate? Machinists like gages that operate like their CNC machines; once it's been programmed, you push a button, and the machine runs, cuts a feature, and is ready for the next part. Gaging should be simple too, requiring as few steps as possible to generate results. If a variety of parts are to be measured on the same gage, it should allow for quick, easy changeovers. Electronic gaging amplifiers, computer-controlled gages (such as surface finish and geometry gages), and even some digital indicators are programmable, so that the user only has to select the proper program and push a button in order to perform the measurements appropriate to a particular part.

No matter how well protected it is against contamination, if a gage is used on dirty parts, or in a dirty environment, it will get dirty. At the end of every shift, wipe down the master and place it in its storage box. Wipe down the gage, and inspect it for loose parts: contacts, reference surfaces, locking knobs, posts, arms, etc. Do this every day, and you will probably prolong its life by years or at least, you'll make it easier for the calibration department to check it out and verify its operation.

Pretend for a moment that you've just installed a new planer in your basement woodshop. Glowing with pride, you set it up, adjust it, and then, just for fun, you make a big pile of shavings. And next? I'll bet you clean it off carefully, maybe oil the iron posts, and promise yourself that you'll always follow the recommended maintenance procedures.

Not a woodworker? Then you probably treat your golf clubs, boat, Harley, or flower-arranging equipment with the same pride of ownership. So why not your gages, which are far more precise than any of these, and deserve far more attention.

Tired Of Bickering Over Part Specs? Standardize The Measurement Process

How many times have you heard an assembly operator complain that incoming parts are consistently out of spec? How many times have you heard the parts people trash assembly folks for not knowing how to use their measurement tools? They were shipped good parts and now they measure badly. What's going on here? Where is the problem?

During the manufacturing cycle for a part or product, many people will look at the part to determine whether it meets the specification. Typically, these could include the machinist producing the part, a QC person, an incoming inspector at the company using the part, and finally another inspector who may be responsible for evaluating the manufactured part's performance within an assembly.

With this many inspection processes, it's very unlikely that they will all be similar, let alone use the same gaging equipment. Even if skilled craftsmen at each inspection point follow their particular measurement processes to the letter, there will, at times, be unsettling disparities in measurement results.

Let's look at a very simple part, a cylinder 1" long × 5" diameter having a tolerance of ±0.0005". How many ways could we measure this part? Here are some of the most popular: micrometer, digital caliper, snap gage, bench stand with anvil and dial indicator, air fork, light curtain, optical comparator, special fixture w/two opposing gage heads and electronic amplifier or, even, a vee with a digital indicator.

Add just a little form or surface finish variation in the part and it's very likely that each one of these inspection systems mentioned above would produce a different result. Even with gages in top condition, there will be slight, if not major, differences. Simply the type of gaging being used and how they contact the part may cause this.

Suppose one company is using an air gage to measure the roundness of a part and another company is using a snap gage. In addition, let's suppose that the part has a very coarse surface finish greater than 50 RA. In this case there will rarely be a correlation between measurement results. This is because the air jet tends to average the peaks and valleys in the finish, while the hard contacts of the snap gage will ride on the peaks. This situation is a disagreement waiting to happen.

What if one company is using a two-point contact gage and the other is using a gage with three contact points? Will the results be comparable? Not if the part has an odd number of lobes. It is an interesting quirk of geometry that a two-contact point gage, as it is rotated around an odd-lobed part will always see the parts diameter. Another disparity between inspection systems is about to happen. On the other hand if a part has an even number of lobes, both gages will deliver comparable results.

One more: What if the part isn't quite straight and one inspector measures with two direct contact points and the other uses a snap gage which makes two "line" contacts with the part? As you rotate the contact point gage, it follows accurately around the part's diameter but the line contact tool will interpret the part's out-of-straightness as out-of-roundness.

There are certainly dozens, if not hundreds, of variations on this theme, but you should get the point by now. While a variety of tools may be used to measure a given dimension, a disparity in measurement processes up- and downstream from your inspection point will, sooner or later, cause unneeded rejection and delays in the acceptance of the part.

If this tune is all too familiar and you don't want to hear it any more, get together with your suppliers and customers and standardize your measurement tools and processes particularly for critical dimensions. It may seem like a little extra work, but in the long run, everybody will benefit.

Starting From Zero

Writing '00 instead of '99 reminds me of the importance of zeroing out the measuring instrument or gage before starting to make a measurement. Zeroing sets a reference point from which all subsequent measurements are made. If a gage has been allowed to drift from zero, it will introduce error into the measurement process. So it's important to "Think Zero," and think it often.

Why do gages shift their zero point? There are probably as many reasons as there are types of gages. But the top reasons include: wear, temperature effects, loose gaging components and dirt. It's important to check zero as often as needed, so that you feel comfortable with the measurement process.

Basic instruments like micrometers or calipers use their scales as the reference. With a vernier micrometer, verifying the reference point is straightforward. Close the contacts together and read the vernier scale. The scale should indicate zero. If it doesn't, you can be sure that every measurement from then on is going to be off by the amount indicated on the scale.

There are two things you can do to correct the problem. The lazy man's way out is to add or subtract this offset with every measurement. This is a trap that may cause a lot of grief the first time you forget to apply the offset.

The best thing to do is correct this disaster waiting to happen by adjusting the micrometer to make it read zero with the contacts closed. Most micrometers, both friction and ratchet drive types, provide instructions to adjust this zero point. Follow the instructions carefully and you will have your micrometer zeroed out in a matter of minutes.

Similarly, a vernier or dial caliper can be checked by bringing the contacts together and holding the jaws up to a light. You should not see light passing through the jaw surfaces. Look for gaps or taper conditions that indicate a worn jaw. If the jaw passes inspection, check for the zero readings. On a vernier caliper you will need to read the lines, while on a dial caliper the indicator should read zero. Both can be adjusted to read zero.

Digital hand tools are easy to zero. Close the jaws and press the zero button. That's all there is to it. The instrument does this important little task electronically.

Comparison type measurement hand tools such as snap gages, gage stands, bore gages, etc., may be a little trickier, but they also need to be zeroed regularly. The process is slightly different, but the end result is the same. With this type of gage the zero point is actually a reference dimension to which dimensions on the parts will be compared. An ID, for example, will be shown to be greater or less than the zero (reference dimension).

For dial indicators, the method is to mechanically adjust the dial indicator on the master so that it is in its midrange, and lock it firmly into place. Then loosen the bezel clamp and turn the dial so that the indicator hand lines up with the zero on the dial.

Something else to remember is that setting to zero this once does not end the process. Take the master out and replace it a number of times in the gage, and check for zero again. A bit of dirt may have been introduced in the initial setting and repeating the process a number of times will help instill confidence in the set-up. Usually you want to have a dial indicator reading repeat its zero setting to within a half of a grad or so, or a digital readout should be to within one count. This varies a little depending on the gage and the resolution, but in any case we are assuming that the gage has already been checked for repeatability performance.

You should also know that there are instances where you may want to set your gage to a value other than zero. This makes it possible to correct for a known error in the master or to use a different size master for measuring the part. For example, if the master is +.0002" larger than the nominal dimension for the part, you would set the dial on the indicator to +.0002" instead of zero. Now if you have a perfect part, the gage would read 0.0" when the part is measured. Electronic dial indicators and amplifiers, in addition to zeroing buttons, usually have master deviation functions to do the same type of correction.

Zeroing the gage is the very foundation of good measurement practice, but we know from experience that most gages are not zeroed often enough. If too many measurements have been made or too much time has elapsed since the gage has been zeroed, measurements will all be biased by zero shift. An extreme solution would be to zero, measure, and zero for every part. This would be overdoing it in most cases. On the other hand, zeroing once a day is probably too little. Generally speaking, once an hour is just about right, but the application itself should dictate the zeroing frequency.

■ ■ ■

It Don't Mean A Thing If It's Got That Spring
Or What To Do When Your Fixture Isn't As Fixed As You Thought

Believe it or not, one of the most overlooked problems in qualifying gages is unaccounted for deflection of the fixture due to the force of the probe on the part. Who would have guessed? After all, fixtures are used to provide stability.

Most fixtures are made of several component parts, and are, in essence, a variation of the well-known C-frame. If the user is aware of some common problems that can affect the use of the C-frame and other fixture designs, he can quite easily detect and eliminate possible error sources like deflection.

All materials, regardless of their hardness, have some degree of elasticity. That also applies to the frames we use to fixture our parts for gaging. Small as it may be, this elasticity is a real and vital consideration in a precision gaging setup. Even the slightest pressures will cause some deflection of the frame. If the deflection is great enough, it will throw off the calibrated accuracy of the gage.

There are several possible solutions. You can: (1) increase the spring rate (i.e., stiffness) of the gage frame to the point where deflection is no longer great enough to affect calibration; or (2) reduce the spring rate of the indicating system until the deflection of the frame becomes insignificant in comparison; or, lastly, (3) compensate for the deflection. While it is possible to compensate for deflection with a reasonable degree of accuracy, your best bet is to take the problem completely out of play with one of the first two choices.

The tendency for an object to deflect is known as the "spring rate." It is the ratio of the load applied to the fixture component (expressed in pounds) to the resulting deflection (expressed in inches). So the higher the spring rate, the less the frame will deflect under a given load. The size of the part to be gaged also figures heavily in whether or not spring rate will be a large or small problem. Large frames designed to accommodate sizeable workpieces are much more susceptible to error caused by frame deflection than are the ones for small pieces. It's a matter of leverage.

As a rule of thumb, the spring rate of the fixture should be at least 100 times greater than the spring rate of the indicator. Fortunately, there is an easy way to test this without doing a lot of math. With a workpiece in your gaging system and the indicator on zero, place a known weight (e.g., one pound) on the arm of the frame at the centerline of the indicator's spindle. The deflection of the frame will be shown on the indicator of the gage. Let's say, for example, the deflection is .004". By applying the 100:1 rule, we know that the load on our probe to make a .004" measurement should not exceed about 100th of a pound or approximately 4.5 grams. Otherwise, deflection of the fixture may alter the result unacceptably.

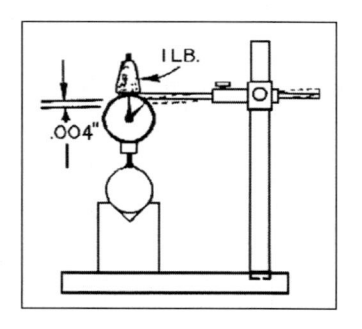

Next, we find out how much load it actually takes to move our gage .004" during measurement. If you haven't removed the one-pound weight, do that. Using a dynamometer, place the lever underneath the contact point of the measuring indicator (dial indicator, electronic probe, etc.). Zero out the indicator and then using the dynamometer apply pressure to the contact until you make the indicator move to 0.004" and note the reading on the dynamometer. If the force is less then 4.5 grams then you're home free.

If the ratio is smaller than 100:1, try making the fixture more rigid or reducing the gaging force of the indicator by going to a lighter force spring. As a last resort, you can introduce a compensation factor to all indicator readings. For example, if the spring rate ratio of the fixture to the gage is as low as ten to one, you may try multiplying the indicator reading by approximately 110 percent to approximate the proper answer.

This added 10 percent will compensate for the fact that one out of every 10 units of part dimension away from zero or nominal results in the spring force of the indicator deflecting the frame arrangement rather than moving the indicator mechanism. It should be understood that this formula for an acceptable spring rate ratio applies only to comparative gaging, i.e., gaging in which the instrument was initially zeroed with a master of the same size as the part to be checked. Spring rates do change with displacement. Also, this process also assumes that there is no part deflection as a result of the measuring forces.

The concept is more difficult to explain than it is to test for. Work through the steps for one gaging setup and you will have mastered a valuable skill to use whenever you suspect that fixture deflection may be causing a problem.

Nine Enemies Of Precision Gaging

In some plants, metal parts are made accurate to 0.01 inch. In other plants, there are products that cannot tolerate size differences of even a few millionths of an inch. Making parts to either tolerance range is impossible without accurate gaging. However, accurate gaging is impossible if liberties are taken with the design, handling and maintenance of precision measuring instruments.

Understanding the following nine major enemies of precision gaging will help defend your measurements against inaccuracy.

Wear. This is the enemy that is most often ignored. For example, linear measurements are usually made by contact between gaging and workpiece surfaces. The gage wears a little each time it's used, and inaccuracy grows by attrition. Wear also deforms gage contacts and flattens spherical contacts, producing discrepancies. The best therapy for gage wear is systematic checking and calibration against accurate masters.

Dirt. Many measurement errors can be traced to someone's grubby hands. Those who measure in millionths of an inch should exceed even surgical standards of cleanliness. This applies especially to people who can't seem to wring gage blocks together without using what is known as wrist oil, a mixture of pore effluent, skin particles, grit, oil and coolant, that coats gaging surfaces with a cement-like sludge ranging from 0.00005 to 0.0005 inch in height. Abrasion from dirt also speeds wear and causes internal looseness.

Looseness. The average user of gages tends to make sure the relevant screws, nuts and clamps are secure. However, internal looseness caused by wear may fool the user. For example, sometimes gage platens and bracket arms creep or a workpiece doesn't settle firmly into place. The key to diagnosing looseness is measurement repetition. If the same reading doesn't come up twice, or if an indicating meter hand flutters irresolutely, then looseness is the likely culprit.

Deflection. Ever present and active, deflection is never seen or felt except by special means. Isaac Newton described deflection in his third law of motion, which states that for every action, there is an equal and opposite reaction. Visualize pushing a cylinder into a gage. Although the contacts separate to accept it, the internal clamping force of the spindle acts equally against the frame, thus causing it to deflect slightly. What is being measured—the workpiece, the frame deflection or both?

Gaging Pressure. This force must be heavy enough to have unwavering authority, but not so heavy as to deform the workpiece. Pressure errors almost always stem from too much rather than too little force.

Temperature. Everyone agrees that a workpiece is bigger when it's hot. Any action taken to alleviate this usually involves cooling the part too much at the nearest drinking fountain. There should be a big flashing sign in every precision gaging area that reads: "Keep the temperatures of the workpiece, gage and master the same."

There is also a tendency to put measurement equipment in locations that cripple their effectiveness. Favorite spots are next to radiators, in the direct rays of the afternoon sun or near a door that's opened and shut 50 times on subzero days.

Vibration. There are people who put a "millionth" comparator near an aisle used by fork trucks. Others sit them next to air compressors or thumping punch presses. The moral is: Do precision work where your comparator won't get the jitters.

Geometry. Measurement must be square to the axis. This is elementary, almost to absurdity. Nevertheless, it points out a major source of error. Whether the instrument is a hand "mike" or an interferometer, many operators persist in cocking the workpiece or cramping the gage just enough to get a wrong answer.

Approximation. A look at any mechanical micrometer reading shows where this enemy lurks. Perhaps it reads 0.494 inch—and a little more. What's your guess on the "little more"—0.4942, 0.4943, 0.4944, or 0.4945 inch? Do you use this as if it were the true reading? Approximation probably causes more wasted time and money in rejections, salvaging, disputes and correspondence than all other measurement errors combined. The usual cure is to get an instrument with higher magnification, or one with an accurate scale subdivided more closely. Another solution is to switch to a digital readout.

There are other known causes of gaging error, and there are still more to be discovered. However, the firm that tackles this list will have taken a big step toward greater precision and accuracy.

■ ■ ■

Say Hello To Mr. Abbé, Mr. Hooke, And Mr. Hertz
(Although You've Probably Met Them Before)

Whenever you use a hand tool or precision gage, you should be aware of typical pitfalls that prevent good gage performance. These include measurement errors resulting from environmental conditions (dirt & temperature), loose and/or worn gage parts, along with operator misuse. But there are other gaging pitfalls lying in wait, which must be considered, both as part of good gage design and good measuring procedure. These issues become particularly important as the need for gage performance increases. They are typified by our three friends, Mr. Abbé, Mr. Hooke, and Mr. Hertz, and most likely sit with you while you are performing your gaging process. Let me introduce you.

Meet Mr. Abbé, a noted optical designer. His principle states: maximum accuracy may be obtained when the reference scale and the workpiece being measured are aligned in the same measurement. This is the case when using a standard set of micrometers: the measuring scale, (i.e., the micrometer barrel or digital scale) is in line with the part and the reference contact. However, in the case of a caliper gage, this is not true.

With this type of gage, the measuring scale is below the contacts, and if there is any angular moment to the jaws during measurement, Mr. Abbé says hello (Fig. 1).

Another type of gage where the measurement scale has often not been in line with the part is the horizontal length machine. Notice that on many of today's newly designed horizontal length measuring machines, great effort has been made to place the scale either in line or as close to the measuring line as possible.

Also, say hello to Mr. Hooke, a physicist who determined that the amount a spring stretches varies directly with the amount of force applied to it. Now, you may say there are no springs in my micrometer, but in reality, the frame of the gage actually acts like a spring. Since springs are sometimes made of steel, steel gaging frames can act like springs. Whenever you turn the barrel of a micrometer down onto the part (Fig. 2), you are applying force to the part and the reference contact. There is some "spring" action taking place in the frame. This is one of the reasons, as we noted previously, for the friction or ratchet drive used in the micrometer barrel. By employing a constant measuring force, we always incorporate the same "spring" of the frame into our measurement, and improve repeatability. This ratchet in the micrometer is a reminder that Mr. Hooke is still around.

And finally, we welcome Mr. Hertz, another physicist who developed a formula that determines the amount of surface deformation within a material's elastic limit (remember the springy steel) when two surfaces are pressed against each other under a certain force.

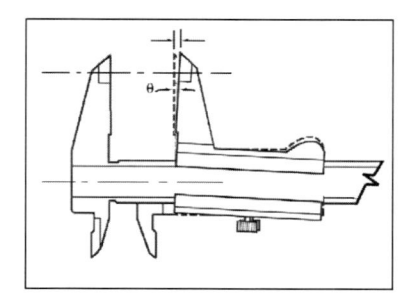

Fig. 1

There are formulas for cylindrical, spherical, and planar surfaces. These formulas are important for determining the deformation of a workpiece caused by the measuring force. The results determined by these formulas are typically small when working with steel, but these small umbers turn out to be important in gage block measurement. They also need to be considered when dealing with compressible materials.

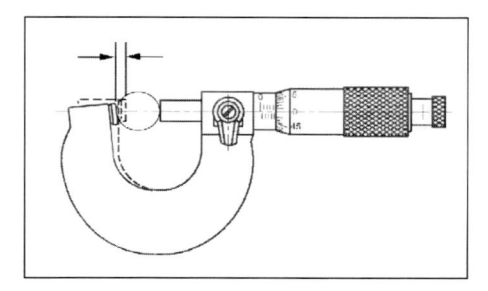

Fig. 2

Section B: Basics of Measurement 55

For example, let's say we have a 0.125" radius diamond contact applying a gaging pressure of 6.4 ounces on a gage block. With a steel gage block, penetration would be approximately 10μ". However with a carbide gage block it would be about 6.6μ". Pretty small stuff—but in the gage block world, the difference can be a major portion of the block's tolerance. And if you know about it, you can deal with it.

In short, when all three of these guys get together and put their forces to work, it's party time! Understanding their principles, and planning for their presence, keeps them under control.

A Different Differential

When we talk about differential gaging, we are usually referring to the process of using two sensing devices and combining the results into one measurement. The measured dimension is the change in the position of the two sensing components.

We've talked before about why differential gaging has some advantages over normal, comparative measurements using a single sensing head against a fixed reference surface. These include the measurement of size without regard to position (see Figure 1). When the two gage heads are in line and in an opposed position, the sensed dimension will be the change in the separation of the two gage tips: in this case, the size of the part.

When we measure in this manner, the staging of the part does not become part of the measurement loop. The platen in a differential system just places the part between transducers. With a single head system, the platen is part of the measurement loop, and therefore flatness and configuration of the platen is critical.

There are available a number of different measuring systems that can provide differential gaging, including air and electronic gaging. Air gaging is probably the most common, since every two-jet air plug and air ring utilizes the differential gaging principle. Most air gages measure backpressure that builds up in the system when the tooling is placed in close proximity to a work-piece: this results in higher air pressure, which the gage comparator converts into a dimensional value. Thus, as the plug fits into the part, it is the combination of both jets that represents the diameter of the part—without regard to where in the part the plug happens to be.

The same is true with electronic probes. In this case though, the probes are combined electronically to provide the differential measurement. Differential gaging using both air and electronic probes is used in a wide variety of applications that include measurement of form on angles and tapers and shafts without regard to the part dimensions. Also, concentricity of two shaft diameters, comparing a work-piece to a master, match gaging, and finally, checking parallelism of a work-piece and its support surface are all applications of this technique.

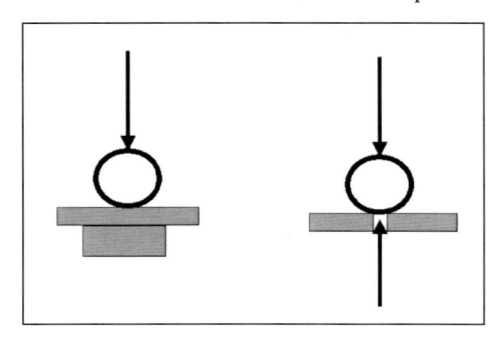

Fig. 1

However, there is another form of differential measurement that is probably not used as often as it could be. This is a mechanical method that provides the same results as air or electronic gage-based differential measurement, but it requires only a single sensing head. It's accomplished by providing a means for the body of the indicator, which becomes the second sensing head, to move along with independent movement of the contact itself. Since the system uses only one sensing head, it can provide very high performance in a space much smaller than when two sensing heads are required.

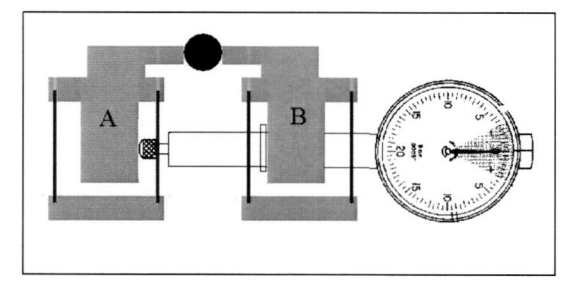

Fig. 2

Figure 2 shows an example of a simulation of a mechanical differential check. In this case we are simulating the measurement of a ball, but it could very well be any other type of length measurement. The method uses friction-free panto-transfer units, but they could very well be any high precision ball slide. By looking at the figure we can see that if we held panto "A" fixed and moved panto "B", the stem of the indicator would move and the amount would be shown on the indicator. On the other hand if panto "B" was held fixed and "A" was allowed to move, the rack of the indicator would be the sensing portion and the motion would also be shown on the indicator. Once both are allowed to move, the combination of both panto transfer units make up the differential measurement, without regard to where the part is located within the range of the measurement heads.

Making use of this simple, but effective method can produce equal and sometimes superior results. This technique can be very handy in situations where there are space limitations and two transducers don't fit into the gaging station.

Another Way To Square It Up, Or Is It 'Perpendicular' It Up?

There are a number of tools available for shop personnel to evaluate the right angle relationship between two surfaces.

The basic machinist square has a number of variations, the most common being the hardened steel square. It is used to check right angles and set up milling and drilling machines. The hardened steel square consists of a thin blade and a thick beam that are set at precise right angles to each other. The square has no scales and is not useful for linear measurements. To evaluate the right angle, the user holds the thick beam on the reference surface and the blade against the side of the part, then looks for light between the blade and the part, or slips feeler stock between the two. These types of squares are usually used on work where tolerances of 0.001" are called out.

The funny thing about these so-called "squares" is that they are not squares, and are actually used to check the right angle relationship between the two surfaces, also known as perpendicularity. Most prints have a call-out requiring a right angle relationship, but one right angle does not make a square. There may also be a call-out for a part to be square, but this refers to the geometric shape of the part. It may be that the reason for calling this simple tool a "square" is that it's too difficult to say, "Hand me that perpendicular." But since this is the language used around the shop, we can keep talking about inspecting for the right angle as a squareness check.

A cylindrical square can be used in a manner similar to the machinist square. By placing the cylinder next to the part and using the same visual or feeler stock check, the operator can get a very good sense of right angle.

There are also a number of other hand tools used to inspect for the square form in a part, including combination squares, linear and digital protractors, and even electronic levels. But if parts are over 8" in length, hand tools cannot cover the range needed and surface plate tools are required. In addition, hand tool methods all rely on the observation and skill of the operator to interpret the angle. None of them provide any empirical data that can be analyzed or used to begin controlling the process.

For large part/surface plate work, one of the best means to inspect for squareness is to use the precise vertical ways that are built into a height gage or master squareness gage. Both of these gages have a precision slide to which a dial or test indicator can be mounted. This allows the indicator to be moved in an accurate, vertical line of travel when both the gage and part are on the surface plate. The advantage of this type of squareness gage over handheld, visual squares is that the dial test indicator allows the operator to read the exact amount of error instead of judging it by eye.

Since both the master squareness and height gages use the same reference surface as the part (i.e., the surface plate), and as the gages themselves provide a precise reference for the vertical axis, both are capable of measuring the perpendicularity of the side compared to the base. On the downside, master squareness gages are not capable of measuring the horizontal axis of the part, and in neither case are the vertical and horizontal readings tied together, so the user has to plot the individual values to come up with a measurement. But if you are really interested in data collection, you'll want to use a motorized electronic height gage. These gages not only allow automated positioning, but also have the capability of using a high-resolution linear encoder for positioning the indicator along the vertical axis, and a similar linear scale for the horizontal axis. Most electronic height gages have dedicated, preprogrammed functions for checking perpendicularity. All the operator needs to do is enter the length of the path to be inspected and the number of readings to be taken along

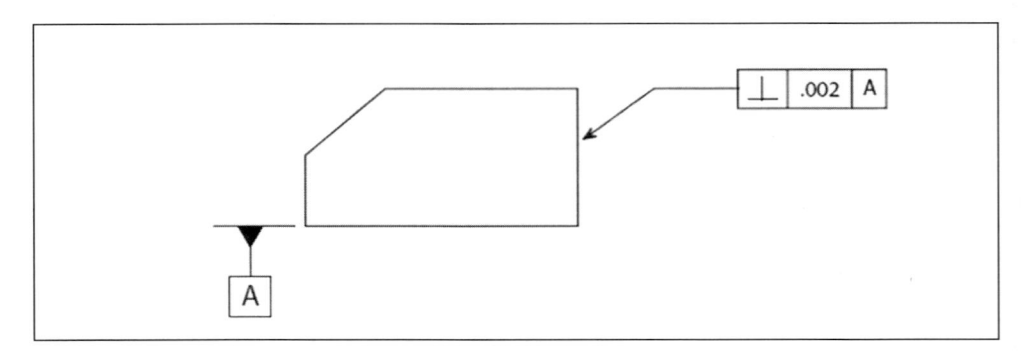

Checks for "squareness" actually measure the right angle relationship between two surfaces, also known as perpendicularity.

that path. The slide can be positioned manually along the path (as with the master squareness gage) to let the gage controller collect the data, or the gage can execute an automated data measurement routine.

Once the measurement cycle is completed, the processor can provide the actual angle measured, the full table of the test part, or even a graph of the part profile. This can be invaluable for large parts where lapping can be performed to fix demonstrated high or low spots. As with any surface plate work, the plate is the reference for both the part and the height gage, therefore a clean plate and a high degree of flatness are essential when making precision checks.

In sum, with the handheld, machinist square measurements, operator influence and visual techniques limit the process to 0.001" tolerance levels. Height gages with high performance digital encoders, long-range measurements, automated gaging routines and computing capabilities can bring surface plate measurements to levels of 0.00005" or better.

■ ■ ■

Where The Rubber Meets The Road – A Probing Look At Probes

The capability of a measuring instrument often comes down to how the contact point interacts with the parts being measured, i.e., the probe or contact. Here are some things to keep in mind when using a contact tip or probe arm as part of your measurement.

Many contact points are made out of a single piece of material. Some, however, consist of ruby or diamond inserts swedged or glued into the tip. Rubies are often used on surface finish probes, while the diamond contacts may be used on millionth class gages or where wear may be deemed extreme. Not infrequently, these ruby or diamond inserts become loose or even fall out. On a surface finish gage, a missing diamond usually will result in the same value being seen no matter what is measured, while a loose contact will typically cause apparently valid, but non-repeatable readings from the gage. A loose contact should be one of the first things checked when this condition is seen.

The material of the contact can itself affect the reading in a number of ways. Certain materials don't mix well together. When measuring aluminum, for example, you should stay away from carbide contacts. Carbide is porous, and aluminum can imbed itself into the contact. Over time this builds up and can produce an offset in the readings. Even though the gage is mastered, the measuring loop has been changed and incorrect readings will result.

Contacts should be inspected for flats and scratches caused by continuous wear. Just as material can build up on contact, it can also be removed. Flats on a spherical contact will produce offsets. Worn areas in a caliper will produce inconsistent readings. Scratches can raise high points on the measuring surface and cause errors. When gages go out for calibration, a complete inspection of the contact is required.

Gages are designed to be used with the correct probe for the application. Sometimes, in an effort to get a gage up and running again, contacts are substituted. This is usually not a good idea, but if you must, use flat contacts when measuring round parts and radius contacts when measuring flat parts. Using the wrong combination will make the measurement very difficult, or if the contacts are not parallel, incorrect.

Probe radius is also important. Some gages require a specific radius for their application in order to meet an industry-wide specification. For millionth measurement, for example, contact penetration is dependent on the geometry of the radius. Changing this will affect the performance of the gage and prevent correction factors from being applied correctly. Always verify that the geometry of your probe meets specifications.

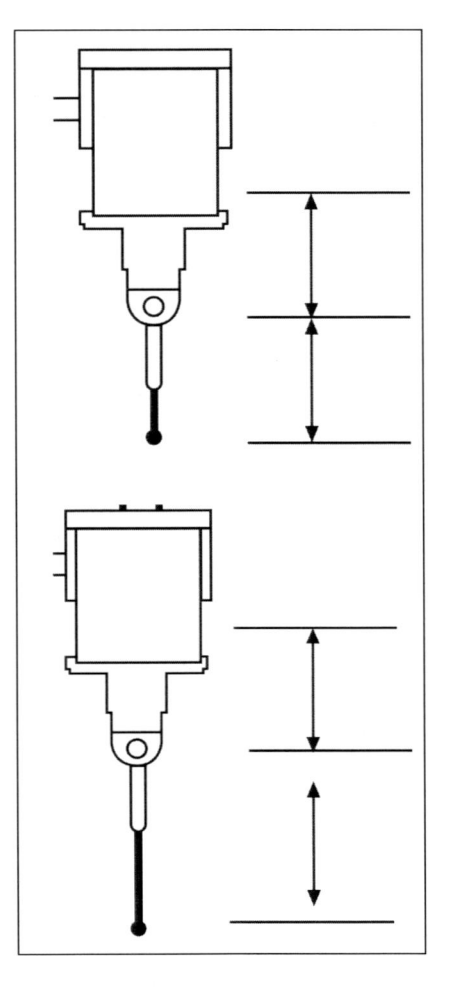

The same is true with surface finish probes. There are specific contact point radii called out as part of the surface finish parameter. Using a 0.0004" instead of a 0.0002" radius probe will provide completely different results.

Lever probes on geometry gages or contour systems can be very long—sometimes up to 10". For contacts this long special designs must be used to make sure they are as stiff as possible. Otherwise, there is a chance of flexure or vibration in the probe becoming part of the measurement. This will demonstrate itself as noisy, unrepeatable readings.

Gage readings generated by lever probes may also need to be adjusted to compensate for their length. Ratios are used to make these calculations, based on probe length, say 2 to 1. However, while the contact may be designed properly, the actual contact itself may not be quite to print and the ratio may have to be adjusted accordingly. To avoid this problem, the measuring system should be calibrated with the probe as part of the measuring loop, and any discrepancies calibrated out of the result.

Contact points react differently under pressure. Specifications for compressible materials show very well-defined characteristics for size, shape and finish. For example, using a .5" flat contact in place of a .125" contact greatly reduces the ounces or pounds per square inch of force on the material and will result in two completely different readings.

Getting closer to your probes and knowing how they are used will definitely improve your gaging performance.

Dimensional Collateral:

Do Two Sines Equal A Cosine?

It is not just a simple irony to say that comparative gages have their greatest accuracy at zero. And it is for this reason—even though such a gage could provide a direct reading measurement—it is always best to use it as a comparator. However, one of the most common sources of error when using a comparative gage over long range is cosine error. If you are concerned with reliable measurement, you need to understand the different ways cosine error can influence a measurement.

Cosine error is most typically seen with test style indicators and lever type electronic probes doing run-out and concentricity checks on shafts and bores, or in engineering and tool making, doing checks of parallelism and alignment of flat faces. With a test style indicator, accuracy is greatest when the axis of the contact point is perpendicular to the measuring direction (Fig. 1). This is seldom the case, however, and as the angle of the contact to the surface increases, the amount of vertical distance encompassed (change in height) also increases. The result is cosine error. Tables can be used to correct for this error as follows, where A is the angle between the probe and the surface of the part.

Angle A	Correction Factor
5°	.996
10°	.985
15°	.965
20°	.940
30°	.866

In circumstances where a larger cosine error exists—i.e., where the angle of the probe is greater than 30°—it may be better to zero the comparator closer to the actual part size. This will minimize the cosine error in the reading. To do this, select a zero master that is closer to the calculated reading than the actual standard size.

In general, the rule is to always try to maintain the probe angle to within +/- 15° in either direction. There are also special contacts available that help minimize this type of error with a special involute shape manufactured into the contact.

Another place where cosine error can have a negative effect is in a standard benchtop comparator. If the axis of the indicator is out of alignment with the line of measurement on the part,

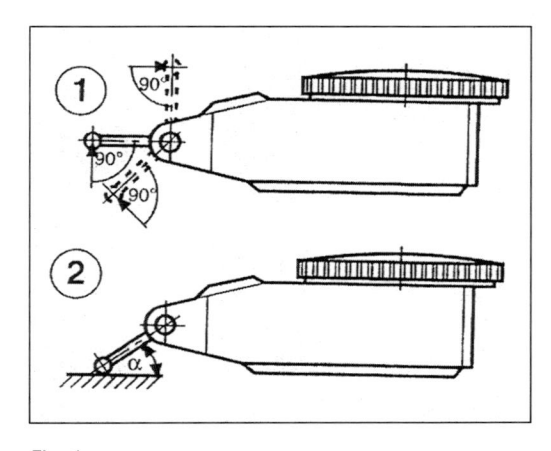

Fig. 1

then a cosine error will result. A one-degree out of alignment condition starts to become noticeable (Fig. 2). If the indicator is set with a 2" master and a part is placed in the gage, a 0.050" deviation will result in a 0.00001" error, as follows:

Change in height = X

$$X = (deviation) \times (\cos 1")$$
$$= (0.050") \times (.99985)$$
$$= 0.04999$$

(deviation) − (change in height) = error
0.050" − 0.04999 = 0.00001"

While this may not be that serious an error for most measurement applications, it does become important when performing gage calibration.

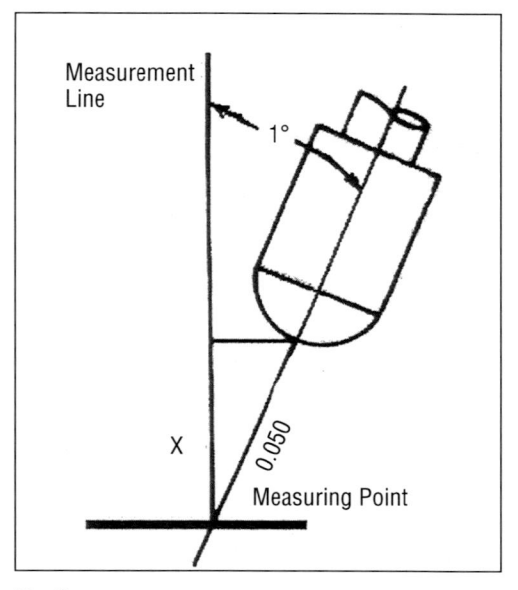

Fig. 2

Although these types of cosine errors seem to get the most recognition, they are the least serious of the errors caused by gage misalignment. In the same example, if a flat contact was used on the gage (rather than the normal radius version, as is most common) the error becomes a function of the radius of the contact surface (Fig. 3). Stated mathematically, this is as follows:

If the misalignment is 1°, the angle of the contact face to the surface is 1°, and the diameter of the contact is 0.250" (radius is 0.125"), then:

$$Error = (radius) \times (\sin 1")$$
$$= (0.125") \times (0.01745)$$
$$= 0.0022"$$

This error is more serious than the previous example of cosine error. Such a misalignment in a micrometer or snap gage would repeat this error at each contact surface. The example also shows why flat contact points should only be used when absolutely necessary. Using a spherical point eliminates part of this error, but not all of it. Since both the contact point and the part being measured are compressible, there is area contact. With area contact comes the potential for sine error.

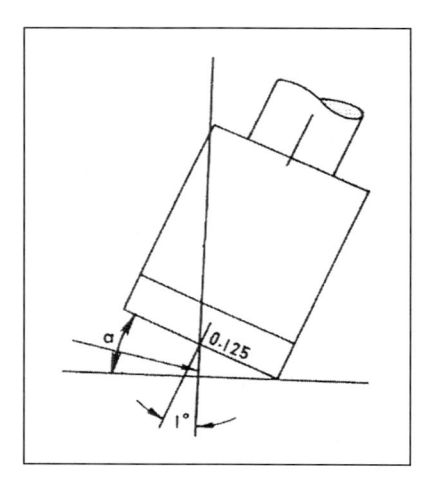

Fig. 3

Getting My Stars Aligned

Good gage design requires certain basic physical characteristics to guarantee reliable performance. A rigid and sound physical design, for instance, helps ensure that operators have as little influence on the measurement as possible. One of the most important of these design principles is that of alignment.

The principle of alignment states that measurement is most accurate when the line or axis of the measurement coincides with the line of the scale or other dimensional reference. Now, in the real world of gages, it is rarely possible to design a gage in which the scale and axis of measurement actually coincide. But the scale and axis should be as close as possible, and definitely in the same plane.

Probably the simplest way to visualize this is in a caliper—whether it be a vernier, digital or dial caliper. They all rely on certain alignments of the jaws to assure correct measurement. While the caliper may not be one of the most highly accurate tools in the metrology tool box, it does typify the types of errors that can be found in much more accurate gages, such as a horizontal measuring machine which needs to be able to perform to microinches.

First, look at the caliper in Figure 1a. It is in perfect condition, with a straight beam, highly machined and straight surfaces, flat jaws that are square to the beam, and a perfect scale. You can see that the line of measurement is pretty well displaced from the line of the scale, but it is in the same plane, and in this case, for any given separation from the jaws, the scale reading will correspond to the separation of the jaws. Now, imagine that the reference scale was mounted to the beam of the caliper incorrectly, so that it was not square to the jaws. This is unrealistic, of course, but it does show how taking the scale out of planar alignment can distort the accuracy of measurement.

A more realistic scenario is shown in Figure 1b, where we have added an amplified curvature to the beam of the caliper. With this type of curvature, the distance between the tips of the jaws is very much less than the distance indicated by the scale reading. However, as we move the contact points of the measurement up on the jaws closer and closer to the scale, the reading gets more reliable, since it is closer to the reference.

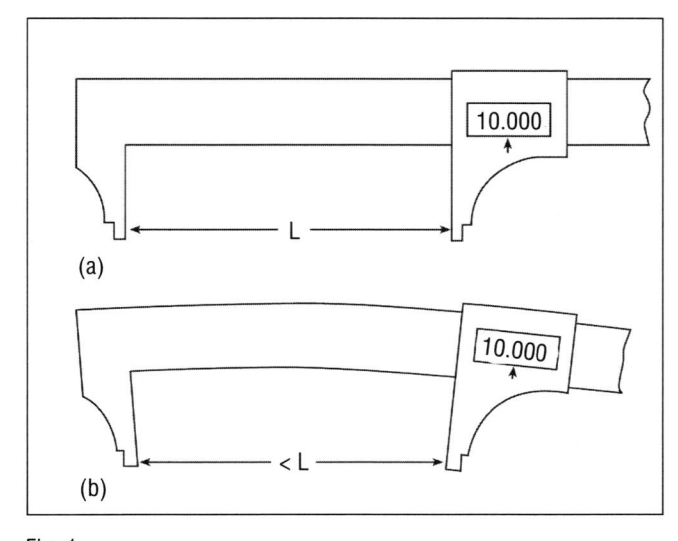

Fig. 1

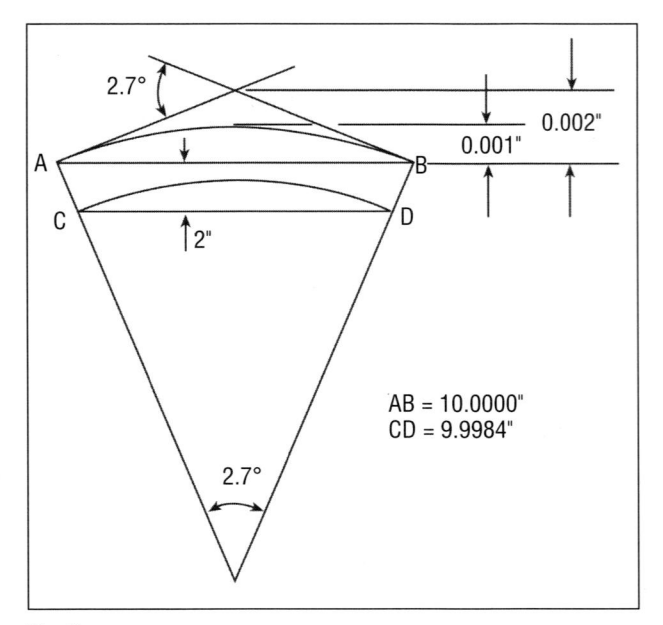

Fig. 2

If we put on our geometry caps, we can think about the actual errors being generated in this example, as shown in Figure 2.

Let's say the curvature of the beam is 0.001" over a 10" beam length. There is a useful rule of thumb we can use to help us figure out this one, which is shown in the diagram. It says, when the height of an arc is small in proportion to the length of a chord—which is always the case in example like this—the apex of the triangle formed by tangents at the ends of the arc is twice the height of the arc.

In our case, since we said the arc height is 0.001", then the height of the formed triangle is 0.002". From this we have some simple right angles to work with, and can calculate the angle between the tangents and the chord to be 2.7°. Now, if we assume the length of the jaws is 2", we can also calculate—I won't make you go through the numbers—that the difference between the tips of the jaws and the reading indicated on the caliper scale is 0.0016".

Most vernier calipers do not have the resolution to see this small an error, but dial and digital units may have the capability of reading it. But what's important here is that with this type of non-alignment condition we are generating real errors. Now, think about this condition in terms of a laboratory universal measuring machine or on a precision jig bore. Here distances could be a lot greater and so could the errors.

Thus, in both hand tools and measuring machines, every effort should be made in the design to make sure all measuring surfaces are aligned to assure best performance.

But what difference does this make to you, the user? How do you tell if your gage is well aligned? With most simple gages, you probably can't, other than to look at the gage with the principle of alignment in mind and decide for yourself how it stacks up. With measuring machines, however, a close look at the specifications for straightness if the ways and squareness will give a good indication of how closely the designers aligned their components.

Plate Gages

4 to 5 You'll Get It Right

Plate gages are a mainstay in the bearing industry, or anywhere that fast, accurate readings of ODs or IDs are needed. You've seen them around: the bench mounted ID/OD comparative gage with the tilting stage plate to set and locate the part being gaged. This basic design, which has been around for over 50 years, is convenient for fast, comparative gaging of flat and relatively thin-walled parts, such as ball and roller bearing rings, where diameter measurements must be made in a plane parallel to at least one of the faces, and sometimes at a particular depth on the ID or OD. Sometimes the location might be the minimum or maximum diameter of the ball bearing race.

The gage consists of a plate that is ground flat, and may incorporate some wear strips on which the part to be gaged is rested. In many cases, however, the plate is no more than a protected surface for the gaging mechanism. Instead of resting the part on the plate, which could cause it to wear and destroy the reference plane, the gaging surface is built into the sensitive and reference contacts of the gage. It is much easier and less costly to replace the contacts on this design, rather than to replace or regrind a reference plate. This design also provides less surface area for dirt or chips to get into the measuring loop and potentially affect measurement results.

There are two types of contact arrangements in these plate gages: a "T" plate design and a "V" plate version. With either version there are movable reference and sensitive contacts that are set close to the diameter to be measured.

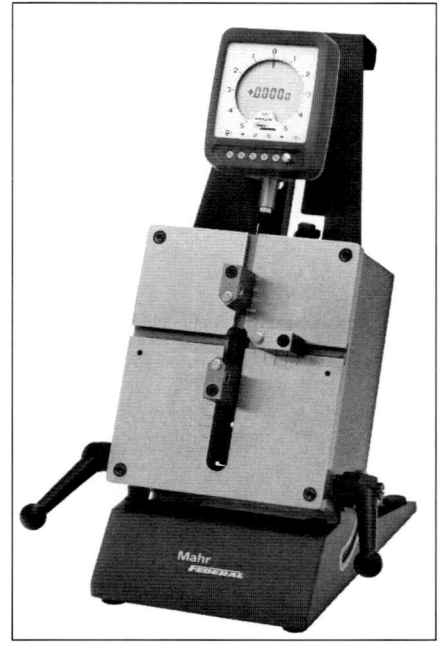

The "T" plate design is the most common and probably the most familiar. Since the reference contact and the sensitive contact are in line, the gaging principle is the same as in a portable snap gage. There is a difference in plate design, however. The contacts used on the plate gage are not flat and parallel as in a snap gage. They are generally curved or "donut"-shaped, which calls for some special consideration. This means that the gage may not necessarily pick up the Max or Min diameter of the part every time. Some slight "swinging" of the part through the contacts is necessary to identify the Min or Max position. The second reference contact on the "T" can help locate the part. However, it should be used to position the part close to the true diameter. It should be set to produce a reading slightly outside of the Min or Max value. Otherwise, if it is set to be exactly on the "zero" diameter, any other position would produce a chord reading and not read the true diameter of the part.

The other contact configuration is the "V" plate design. This design incorporates two reference stops, one at the top of each arm of the "V", that must be adjusted symmetrically to assure that the part is staged on the center plane of the "V". This double stop has a locating effect similar to that of a vee block and provides positive and precise location of the part on the gage. This greatly speeds up the measuring process, taking some of the operator involvement out of the measurement, and is especially useful when the part might contain an odd lobing characteristic from the machining process. However, there is a drawback to this type of contact arrangement. Since the sensitive and reference contacts are not in a direct line, there is not a one-to-one relationship between sensitive contact movement and the diameter. Thus, there are two special considerations that should be borne in mind when using this type of gage. The first is that the angle between the reference contacts determines the multiplier needed to determine the measurement, just like the multiplier used when measuring a diameter on a vee block. In most cases this angle is 60° and the ratio is 4:5. This means that for every four units seen by the indicator, 5 units come out (which is another way of saying the sensitive contact is multiplied by 1.25 to get the correct result).

The other thing to remember about this arrangement is that these configurations work only for comparative readings and cannot be stretched into the "absolute measurement" world. This is because there is a window of accuracy wrapped around the angle setup for the reference contacts. But if the sensitive contact is moved significantly away from or toward the reference contacts—as would probably happen in an absolute measurement scenario—the angle relationship changes. This changes the multiplier needed to get correct results. To correct for this, a scaling multiplier based on the measurement size and the location of the contact would be needed. It could be done, but it's pretty complicated for a bench fixture gage.

Fortunately the user need not worry about these angles, ratios, and long-range measurements. The gages take all this into account, and have been doing so for a long time with proven success.

TIR Versus Concentricity For Coaxiality

One of my favorite customers was having a bad day. "Those boys on the CMM are disputing my gage results," Sue said. According to them, the results from her gages were significantly larger than their CMM readings. More, the process boys liked the CMM results better because they could use them to help control their process. The gage numbers didn't seem to help, so they decided to do it their own way. "Now my whole inspection process is in doubt."

This is not an uncommon problem, I told her, and asked her to give me the details. The part in question was a simple shaft with two diameters, and the check was to ensure coaxiality. The gages Sue was using were designed to measure what the print called out, circular runout, which is the old reliable TIR check. This gives operators a number they can compare against the tolerance to determine if the part is good or bad.

The guys with the CMM decided they would take a little liberty and look at the check

Circular Runout

Concentricity

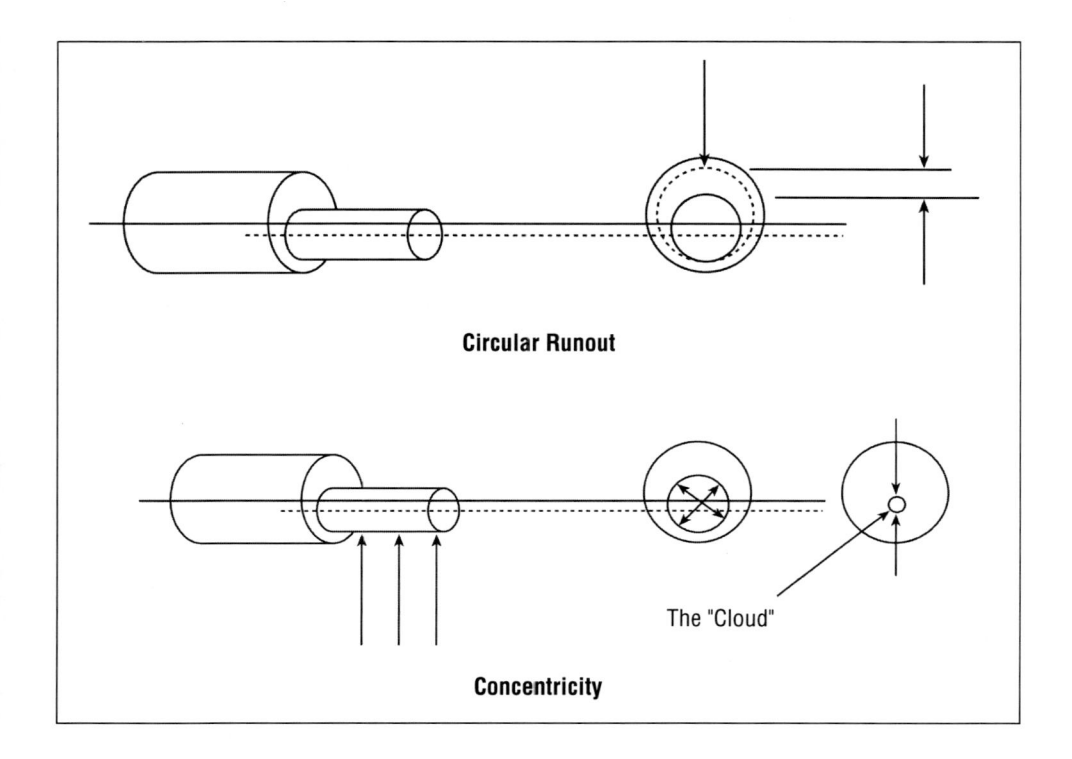

The "Cloud"

with a concentricity function. They felt this was more valuable since it gave the operators an X and Y figure that could be used to offset the machining center to help control the process. This is an interesting approach, but it's also one headed for conflicting results.

Circular runout is a two-dimensional measurement using surfaces to control an axis. The tolerance is applied at any cross-section. When it is used on a surface referenced to a datum axis, as with this part, it will control the total sum of all variations of circularity and coaxiality.

Concentricity is a little more complex. It can be thought of as three dimensional and uses a series of diameters and their midpoints to generate an axis location. This is why the process control guys like it: it gives an X Y location of how the two axes line up.

But why would one be larger than the other? They both start out similarly, with one diameter used as the reference. The part is either held in a Vee type arrangement or a three point clamping system to set up an axis of rotation. The outside diameter and the holder act like mating parts to create the minimum circumscribed cylinder.

In the case of the TIR check, an indicator of some type is used to watch the test diameter as it is rotated. The indicator will see EVERYTHING that the diameter is doing, and the movement of the indicator will include:

- Any misalignment or runout between the reference fixture and the test diameter
- The angular relationship between the two diameters (coning error)
- Any form error in the test diameter

Done correctly, the indicator is placed on the measured diameter, set to zero and the part is rotated through 360 degrees. The measurement must fall within the part diameter tolerance and the runout tolerance. The check should be performed at a number of locations along the axis. This check will not control taper.

With the concentricity check, one diameter is held in a fixture to set up a reference axis (or the CMM may create an axis based on a few diameter touches on the reference). The CMM then takes a series of diameters on the surface being measured and finds the center of each diameter. Any number of diameters can be used, but 4 to 8 is typical. Then the centers of the diameters are plotted to create a set, or cloud of points that can be used the find the mathematical X Y location of the center of the diameter. As long as this cloud of points falls within the specification, the concentricity spec is passed. This check should also be made at a number of different locations (axially) to ensure the whole length is within tolerance.

So why the difference? Simply put, and ignoring any error in form or alignment, the TIR check bases its coaxiality reading on diameters, while the concentricity check calculates radii. It's apples and oranges, feet and yards. However, there is another potential reason for different results as well. The runout check is taking an infinite number of readings while the CMM is taking 4 to 8 diameters. The indicator is apt to see much more form variation than the CMM.

Which one is right? They both are for their particular call-out. If the print specifies runout, then the part should be inspected for runout. If concentricity can be used to help control the process, that's also great for process improvement. But keep these apples and oranges separate and use the right one for the application.

Speeding And Gaging Don't Mix

Manufacturing is under constant pressure to get more productivity out of its process. Machine tool speeds, feed rates, and part positioning all keep getting faster, yielding more parts in less time. On the inspection side, speed is also critical, especially for gaging systems designed for use inside the manufacturing process. A slow gaging system can become a bottleneck if it can't keep up with the parts being spun out of a machine tool.

However, there are physical characteristics inherent in some measuring processes that limit how fast a measurement can be made. There are also characteristics of gaging equipment that limit speed of inspection.

The most common place we see these limitations is in the area of dynamic measurements such as TIR (Total Indicator Runout), Minimum or Maximum dimensions. Other areas include checks for form error such as roundness or surface finish. To do these types of inspections, the part must be rotated against a sensing probe, or the probe is dragged across the part surface.

Runout and roundness measurements are very similar in that they both involve rotating either the part or the gage, and inspecting for a dimensional variation in the surface.

Most electronic transducers have the capability of responding very quickly to changes in variation. However, if the measurements are taken too quickly, the contact point (which has some mass) may tend to bounce when it goes over some high spots on the surface of the part. You can test this yourself. If you have a lever style gage head on a round part, mounted either on a precision rotary table or in a set of vees, and you rotate this against the transducer, you will see the readout device follow the part's variation. If you turn the part slowly you will see every peak and valley as they go by the transducer. If the part is rotated faster, the probe will tend to rise up the peaks, and because of its momentum, keep going a little bit while the part surface starts its downward slope. Speed up the rotation fast enough and the probe may even miss some valleys and skip over some peaks. Most form systems bring the part rotation up to a fixed speed prior to measurement. This serves to best match the response of the probe to the surface speed of the part and provides a constant surface speed for all recorded values.

On surface finish machines the probes are different. These very delicate and sensitive probes are designed for speed. They have very little mass, which helps eliminate tip bounce, and they have very fast electronics to process data very quickly. However, the drive units still travel quite slowly in order to minimize loss of data.

Gaging systems employing digital sensing technology have become more common on the shop floor. These are important tools because they often offer long measuring ranges with very high resolution. This increases their versatility. However, the issue with many digital sensing technologies is how fast the gaging processor reads the probe or probes. Single probe applications yield the best performance because the processor has to monitor only one probe. Add multiple probes and the shared data handling sacrifices the number of readings on the part. If the part is rotating under the probe, there are going to be times when spots on the part do not get measured. On a part rotating very quickly, or one with a large diameter and high surface speed, you are apt to have more of the part not measured than measured. This will certainly affect the ultimate quality of the measurement.

Air gaging also limits the speed of both static and dynamic measurements. Air is a compressible medium and needs time to stabilize. To measure a part with air, two things have to happen: one, if the air jet is open to the atmosphere and then brought against the part, the air lines need to pressurize. This takes about one second per foot of air line. Then, two, as minor changes in the part go by the jet, the air pressure in the line has to restabilize. Thus, if the part is not in position long enough—or if the surface of the part is moving past the open jet too fast—the pressure in the line can't keep up and important part data can be lost. The longer the air line the more time is required for the air reading to stabilize.

How do you determine the best measuring speed? Simple trial and error. Run the gaging station unbearably slow and record your results. Then speed it up a bit and record those readings. Repeat the process until you start to see degradation in results. You'll quickly find out where the limitations are.

If the manufacturing process can't wait for some of these time limiters, however, you will have to work around them. The most common approach is to increase the number of gaging stations. But be wary of faster gaging. If you seem to be getting better results than others who are using a slower process, it might be because you're missing something.

Section C
Surface

Start To Finish

When an engineer includes a surface finish spec on a print, the intent is usually not just to make the part look good. Surface finish affects how a part will fit, wear, reflect light, transmit heat, distribute lubrication and accept coatings. The finish should be determined by the part's function: you want a surface that fulfills the engineering requirements of the application, without wasting time and effort on a higher quality finish than is necessary. (In fact, many applications do better with a certain amount of "texture," and too fine a finish can be as bad as too coarse.)

Thirty years ago, when most dimensional tolerances were measured in thousandths of an inch, the difference of a few millionths in surface finish was often irrelevant. Now that tolerances of "tenths" or even tens of millionths are commonplace, variations in surface finish represent a sizable percentage of the total error budget. Note the following example:

The maximum peak-to-valley height on a surface is usually four or five times greater than the average surface finish, as measured by the R_a method. A part with an R_a value of 16µ", therefore, probably has a peak-to-valley height of 64µ" or greater. If you're trying to meet a dimensional spec of .0001", the 16µ" finish represents over half the allowable tolerance.

As shown in Figure 1, surface finish—also known as profile—is composed of two components: waviness and roughness. Waviness, or longer-wavelength variation, is caused by macro-type influences, like worn spindle bearings, or vibration from other equipment on the shop floor. Roughness—the short-wavelength pattern of tool marks from grinding, milling or other machining processes—is influenced by the condition and quality of the tooling. Both can be influenced by the operator's choice of feed rate and depth of cut.

Although fingernail scratch-pads may provide a usable guide to finish, they can't meet the modern requirements of documentation and traceability. Hence the increasing importance of surface finish gages. As shown in Figure 2, there are two basic varieties: skid-type, or averaging systems, and skidless, or profiling systems. Skid gages have a hinged probe assembly, with the probe riding next to a relatively broad skid that also contacts the workpiece. The skid tends to filter out waviness, so the probe measures only short-wavelength variations. A skid gage has a dial or LCD readout to display the measurement as a single numerical value.

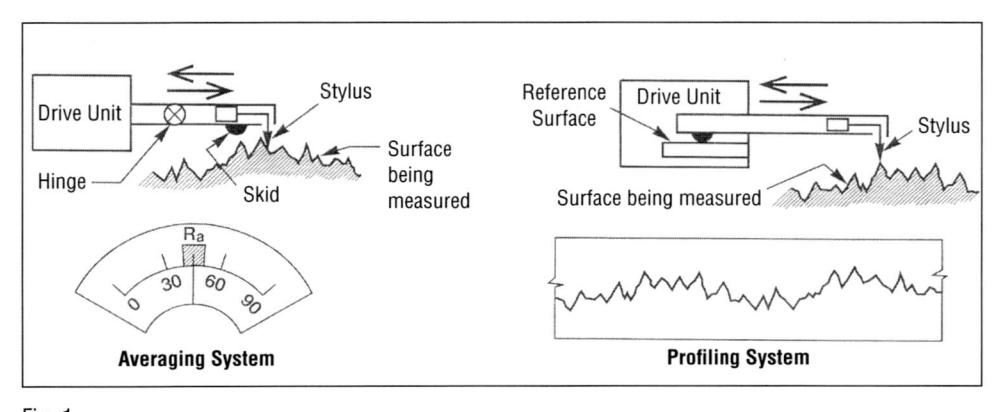

Fig. 1

Skidless gages incorporate a smooth, flat internal surface as a reference, so the probe can respond to waviness as well as roughness. In order to allow separate analysis of long- and short-wavelength variations, profiling gages usually generate a chart (on paper or on a computer screen) rather than a single numerical result.

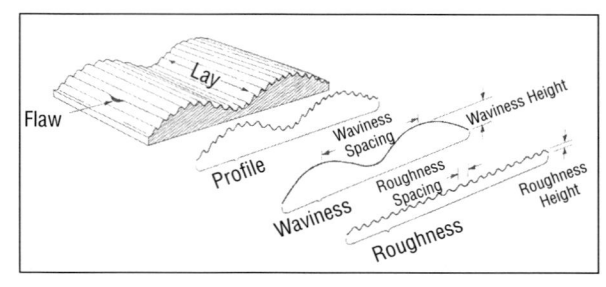

Fig. 2

Every application reacts differently to different combinations of roughness and waviness, and industry has responded by creating more than 100 different formulae with which to calculate surface finish parameters from the same measurement data. Many of these are very application-specific, and most shops are able to confine their measurements to a half-dozen parameters or so. In almost all cases, measurements are presented in microinch or micron units.

R_a is the most widely used parameter, because it provides an arithmetic average of surface irregularities measured from a mean line that lies somewhere between the highest and lowest points on a given cut-off length. A slightly more sophisticated variant, R_q, uses a root mean square calculation to find geometric average roughness—an averaged average, if you will. Both of these, however, tend to minimize the influence of surface anomalies like burrs or scratches. If such factors are critical to the application, R_{max}, R_y, R_t, and R_{tm} all calculate roughness as a function of maximum peak-to-valley height. Also useful is R_z—the "ten-point height" parameter—which calculates the average of ten maximum peak-to-valley differences within the sampling range.

If surface finish is called out on a drawing but not otherwise specified, it is standard practice to assume R_a. But no single parameter is best for all types of parts, and many applications are best served by using two or more parameters: for example, R_a (average roughness) in combination with R_{max} (maximum roughness) may provide a good general idea of the part's performance, and alert QA to the presence of potentially damaging surface anomalies.

Surface finish is not simply a challenge to meet: it represents an opportunity as well. In some cases, if you can maintain good control over surface finish, you may be able to safely reduce precision in other areas.

■ ■ ■

R_x For R_a Measurements

R_a, or average roughness, is the most commonly specified parameter for surface finish measurements. Because it describes the arithmetic average deviation of a surface from a mean line, R_a provides a good general guide for part performance over a wide range of applications. But, as can be expected of anything intended for general-purpose use, R_a has numerous limitations when applications are highly specific, or when small details of surface finish can make a big difference in part performance.

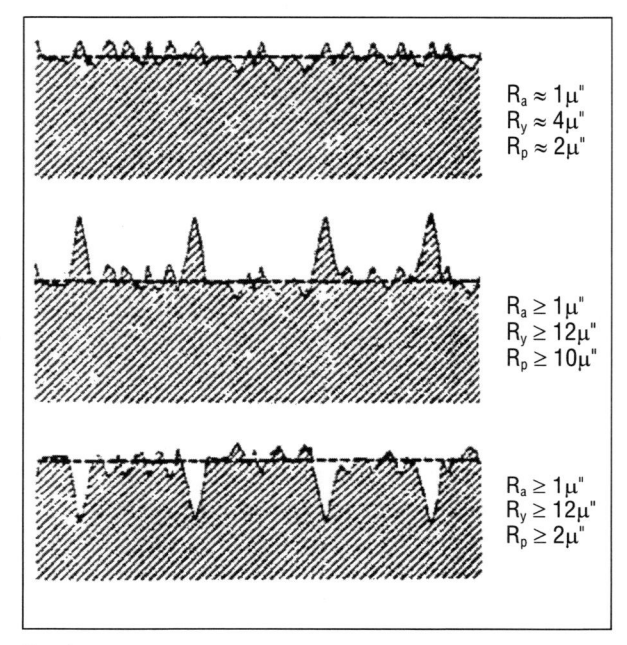

Fig. 1

$R_a \approx 1\mu"$
$R_y \approx 4\mu"$
$R_p \approx 2\mu"$

$R_a \geq 1\mu"$
$R_y \geq 12\mu"$
$R_p \geq 10\mu"$

$R_a \geq 1\mu"$
$R_y \geq 12\mu"$
$R_p \geq 2\mu"$

The key to specifying and using R_a measurements successfully is understanding how average roughness relates to surface finish in general, and the relationship between the machining process and the profile.

As shown in Figure 1, surfaces with different profiles can have the same R_a value, and these differences might be critical in certain applications. The surface shown in the middle trace, if used in a relative-motion application such as a rotating shaft, might score bearing surfaces and cause bearing failure. A part with scratches in its surface, as illustrated in the bottom trace, might fracture prematurely under sheer stress. Clearly, different roughness parameters are required to ensure that the finish is appropriate to the application. In the above examples, the R_p parameter (peak height) could be used to indicate and guard against the condition in the middle trace. To determine maximum scratch depth, as in the bottom trace, one could subtract the R_p value from the R_y (maximum peak to valley height) value.

Engineers and quality organizations who do not understand roughness measurements sometimes specify extremely tight R_a values in an attempt to guard against occasional scratches or peaks. This is an uneconomic approach to quality. We have seen one ball manufacturer who was able to substantially undercut the competition for an aerospace bearing contract by showing the end-user how a looser R_a spec, in combination with control over the R_p parameter, could produce bearing life equal to that achieved with the existing, tighter R_a spec without control over peak height. The ball manufacturer understood that it was the peaks on the balls, not their average roughness, that were principally responsible for scoring the races, and he found it much cheaper to knock the peaks off than to meet the high-tolerance R_a spec that had been put in place.

Different machining processes naturally generate different tool patterns. The roughness produced by grinding, for

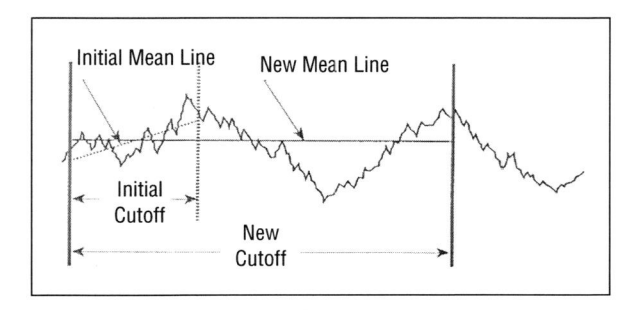

Fig. 2

example, is generally of a shorter wavelength than that left by turning. Milling leaves even longer wavelength patterns, though not as long as those produced by single-point boring. (Note that the wavelength we're referring to here is the spacing of the individual tool marks, not the waviness component of surface profile.)

When performing an R_a measurement, it is essential to choose a cutoff length appropriate to the process. The cutoff length should be short enough so that the measurement will not be influenced by waviness. On the other hand, it must not be so short that only a portion of a tool mark is measured, as shown in Figure 2. A cutoff long enough to include five complete sets of tool marks is desirable to obtain a good average roughness measurement.

Surface finish gages of the simplest type, that only measure R_a, aren't much help in determining whether you've got a peak-and-valley problem, or what the proper cutoff length should be. It may be necessary to perform a complete surface finish analysis, including a look at the waviness component, to get a full understanding of the profile. With that in hand, however, straightforward R_a measurements may be all you need to maintain control over your process.

■ ■ ■

Look Into My Stylii: Care Of Surface Finish Gage Contacts

Proper care of contact points is one of the basic considerations in gaging. Whether you're using a simple indicator gage or a sophisticated surface finish instrument, much depends upon the condition of the sensitive contact point, which is the interface between the gage and the workpiece.

Dial indicator contacts are easy to inspect. They're big enough to get your fingers around, and you can check them visually for wear, damage, or contamination.

The tiny contacts on surface finish gages are quite another matter. Compared to blunt dial indicator contacts, a surface finish gage has a fine "stylus" point, enabling it to follow the texture of the surface, but being of a form and dimension so as to avoid scratching the part. Gaging force from a dial indicator is typically 120 grams/1.2 N; in contrast, force from a surface finish gage is light, usually between 100 mg and 1,500 mg (1 mN to 15 mN). But because the stylus is "dragged" across the surface of the workpiece, it is just as subject to wear and damage as a dial indicator contact.

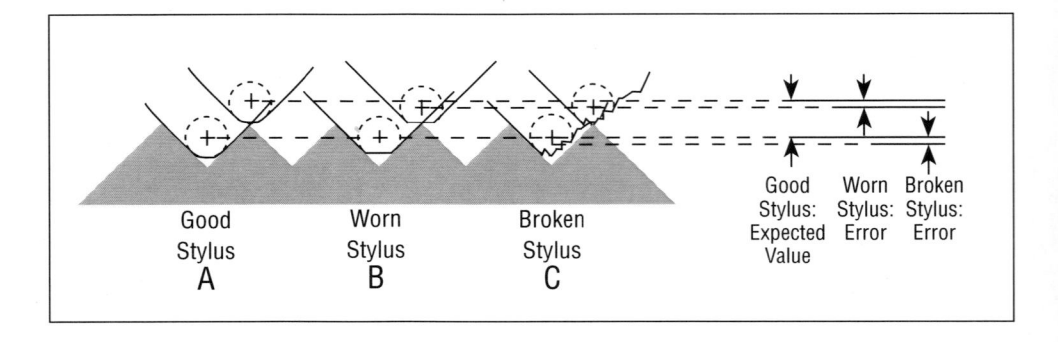

Surface finish contacts are made of diamond, and are conical in form, having a 60° or 90° included angle and a spherical vertex with a radius of 0.0004", 0.0002", or 80 microinches (10 micrometers, 5 micrometers or 2 micrometers). Criteria for choosing between these options are outlined in ASME B46.1 - 1995 paragraph 4.4.5.1.

Even with its small radius, a stylus in perfect condition may be too broad to reach the bottom of the small "valleys" on the part surface, as shown in "A" in the diagram. This inherent "error" is accounted for by calibration, however.

Wear or damage to the stylus will affect measurements. An evenly worn contact will typically under-report the distance between surface peaks and valleys, as shown in "B". A broken stylus may under-report surface variation, or may over-report it, as shown in "C," depending on the nature of the break and the part surface. On occasion, the break may be so sharp that the stylus could scratch the part: a pretty clear indication that the stylus needs replacement.

Paragraph 11.7.2 of the B46.1 standard describes several more practical methods by which stylus condition may be checked. Because the stylus radius is too small to see with the naked eye, a hand-held magnifier or a microscope is required for visual inspection. If it is necessary to assess the stylus's condition numerically, however, a stylus check patch or reference specimen, rated for about 20 microinches, is required.

To use a stylus check patch, a benchmark reading must first be established when the stylus is in new, unused condition. First, the gage is calibrated to a certified specimen rated nominally at 125 microinches (3.2 micrometers). Then it is tested against the 20 microinch patch, and the reading recorded for future reference. The test need not read exactly 20 microinches, because these specimens contain considerable inherent error (e.g., the inability of the probe to reach the valley bottom). For this reason, the 20 microinch patch should not be used to calibrate the gage, but it is nevertheless useful for the purpose of stylus inspection.

When it is necessary to check the condition of the stylus, the gage is again calibrated against the 125 microinch specimen. Then the 20 microinch specimen is measured, and the reading compared against the when-new results. If the reading has changed by more than 25 percent, the stylus must be considered worn or damaged.

As noted above, a broken stylus may be sheared off so that it is either too blunt or too sharp. Conceivably, it could reach exactly as deep into the valleys as a new stylus, and rise exactly as high on the peaks, but the surface pattern that it reports will likely be distorted. So any time readings become non-repeatable, or reflect a sudden change in surface finish with no known change in the manufacturing process, a broken stylus is a likely suspect.

Should damage or wear be discovered, there is only one option available: replacement. Some gages feature easy-on/easy-off contact mounting, while others require a more involved procedure. Users should be aware of the variety of contact shapes available from some gage suppliers. These may include extra long lengths, smaller diameters, or special shapes that provide access to features that are otherwise hard to reach. But no matter what contact shape is used, it should be inspected regularly to ensure accurate surface finish measurements.

Measuring Roughness With Buttons And Donuts

The R_a parameter is the most commonly used measurement for surface roughness. Until recently, in fact, it was the only parameter recognized by ANSI, although new ANSI and ISO standards include many different parameters from which to choose. And while these additional parameters are useful in many applications to ensure or enhance functionality, R_a is still included in most specs as a good starting point and a basic benchmark of process consistency.

R_a can be measured with two types of contact gages, which are distinguished from one another by the nature of the probe or contact that traverses over the part's surface. In "skidded" gages, the sensitive, diamond-tipped contact or stylus is contained within a probe, which has a metal skid that rests on the workpiece. Thus, skidded gages use the workpiece itself as the reference surface. This is a relatively simple, inexpensive approach to surface measurement. Skidded gages sell for as little as $1,600, and some are small enough to fit into a shirt pocket.

Skidless gages use an internal precision surface as a reference. This enables skidless gages to be used for measurements of waviness and form parameters, in addition to roughness. The drive unit is larger and more complex, and a computer is required to handle the complex algorithms for numerous parameters. Skidless gages are indispensable for complex surface analysis but, at a cost of ten to twenty times that of skidded systems, they are impractical if R_a is the only parameter required.

Getting back to skidded gages, it is important to look at the design of the skid itself. Some probes have a simple button-like skid, which may be located either in front of, or behind, the stylus. Others have a donut-shaped skid, with the stylus extending through the hole in the middle. In most applications, both types perform equally well, but occasionally, one or the other might be required to obtain accurate results.

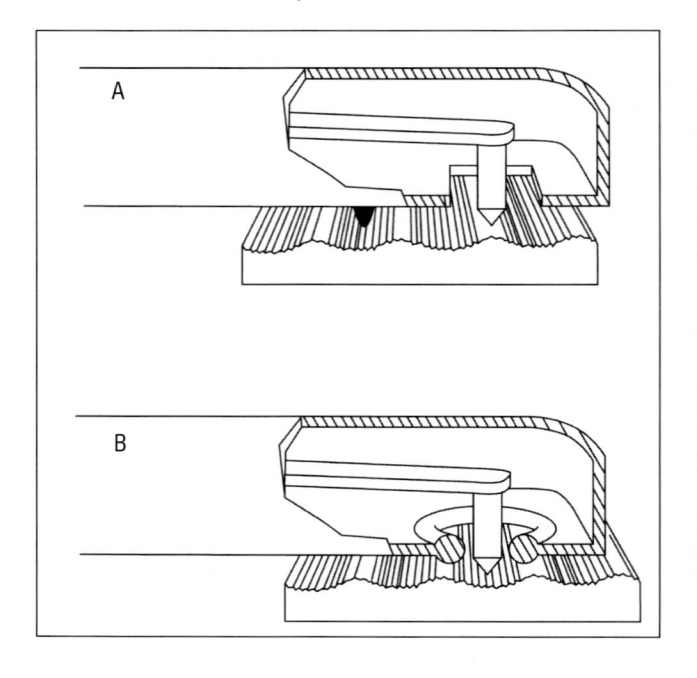

Under high magnification, some workpieces appear to have wavy surfaces of very short wavelength; this is especially so of EDM parts. While the inclination may be to measure these surfaces using a waviness parameter, the pattern is really a tool mark, so a roughness parameter like R_a is required. Surfaces of this type may cause problems for gages with button-type skids. As shown in diagram "A," if the distance between the skid and the contact is roughly half the wavelength of the surface waviness, then the skid and contact will trade places at the tops and bottoms of the waves as the probe traverses the surface. This has the effect of nearly doubling the vertical travel of the contact relative to the reference, which will produce results that may be unreliable or non-repeatable.

The donut-type skid avoids this problem, because it remains at or near the tops of the waves as it traverses, as shown in diagram "B." Thus, the contact's vertical travel is measured against a far more constant reference height.

But because probes with donut-type skids require substantial structure ahead of the stylus, they cannot reach certain features, such as surfaces next to shoulders. Probes with button skids mounted behind the stylus require little or no leading structure, and thus have the advantage of increased access. Special probes with button skids are even available to reach into groove bottoms several millimeters deep.

Some pocket-type roughness gages offer users the ability to switch probes. This can extend the capabilities of the gage, allowing the user to select a probe with a donut skid for use on short-wavelength EDM'd surfaces, and a trailing-button skid for use where access is restricted.

■ ■ ■

Surface Texture From R_a To R_z

The irregularity of a machined surface is the result of the machining process, including the choice of tool, feed and speed of the tool, machine geometry, and environmental conditions. This irregularity consists of high and low spots machined into a surface by the tool bit or a grinding wheel. These peaks and valleys can be measured, and used to define the condition and sometimes the performance of the surface. There are more than 100 ways to measure a surface and analyze the results, but the most common measurement of the mark made by the tool, or the surface texture, is the roughness measurement.

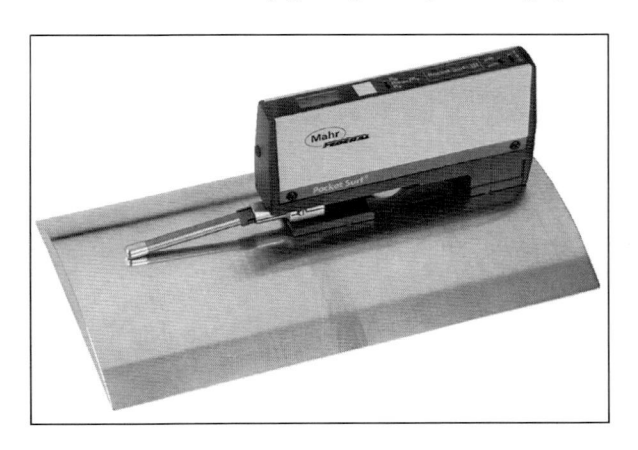

However, there are several different methods of roughness measurement in use today, and the method used on any given part depends largely on where in the world the part is manufactured, and the measurement parameters the manufacturer and the customer prefer to use. It is not uncommon for different parties involved in the production to use different methods for roughness measurement. In this article we will talk about only two of the many methods of roughness measurement, how to convert between these two methods, and how to avoid the problems caused by the inevitable use of more than one roughness measurement.

In North America, the most common parameter for surface texture is Average Roughness (R_a). R_a is calculated by an algorithm that measures the average length between the peaks and valleys and the deviation from the mean line on the entire surface within the sampling length. R_a averages all peaks and valleys of the roughness profile, and then neutralizes the few outlying points so that the extreme points have no significant impact on the final results. It's a simple and effective method for monitoring surface texture and ensuring consistency in measurement of multiple surfaces.

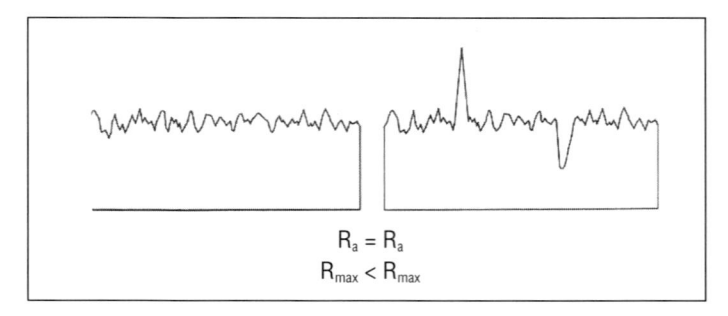

In Europe, the more common parameter for roughness is Mean Roughness depth (R_z). R_z is calculated by measuring the vertical distance from the highest peak to the lowest valley within five sampling lengths, then averaging these distances. R_z averages only the five highest peaks and the five deepest valleys—therefore extremes have a much greater influence on the final value. Over the years the method of calculating R_z has changed, but the symbol R_z has not. As a result, there are three different R_z calculations still in use, and it is very important to know which calculation is being defined before making the measurement.

In today's global economy, machined parts are being made and shipped around the world. As a result, manufacturing and quality control engineers are often forced to decide whether or not to accept a part when the print requirements are not consistent with measurement on the surface gages in the local facility. Some quality control engineers might even assume that if a part is checked and passed using the parameter available, the part would also pass other checks. In these cases, the engineers are assuming a constant correlation, or ratio, exists between different parameters.

If there were no choice but to accept some assumptions, there are rules of thumb that can help clear up the confusion and convert R_a to R_z or R_z to R_a. If the manufacturer specifies and accepts the R_z parameter, but the customer uses the R_a parameter, using a ratio range for R_z-to-R_a = 4-to-1 to 7-to-1 is a safe conversion. However, if R_a is used as an acceptance criteria by the manufacturer, but the customer accepts R_z to evaluate the part, then the conversion ratio would be much higher than 7-to-1, possibly as high as 20-to-1. Keep in mind that the actual shape of the part's profile will have a significant impact on these ratios.

Communication at the outset of the project can avoid most surprises. The approximate and sometimes questionable comparisons can be avoided by developing an understanding of exactly what a parameter on a print means, and how the various parties involved in the production plan to check surface texture.

The best way for those involved in the production to be in agreement on the parameters for measurement is to have capable evaluation equipment in both the manufacturer's and customer's facility, making the same check using the same method. If the manufacturer or the customer uses conversion ratios, then both parties should be aware of the use of the ratio and be comfortable with the ramifications.

Searching For The Perfect, Identical Waves

When two different dimensional measuring systems are used on the same part, one would expect to get pretty much the same results. But in certain cases, this may not always be true. Take for example, a contact gage and a non-contact measuring system such as an air gage. Even though both are measuring the same dimension, there is a significant chance that different results will be obtained because of the way the part's surface interacts with the different sensor types.

The same problem can occur with surface finish gaging itself. Two measuring systems, using what appear to be similar techniques (whether or not they are from the same manufacturer) can sometimes produce very different results. When this happens you need to conduct a comprehensive regimen of process matching to achieve similar results.

First, make sure that the two measuring systems are individually producing repeatable readings on the part they are measuring. Make numerous traces in the same area on the part with both gages to check this. The results of each page should be consistent.

If each gage is producing its best results and they still don't match up, the next thing to do is a settings check. While surface gaging may appear pretty simple—drag a probe over a surface and get some results—it really isn't that easy! A complex combination of various measurement settings and data calculations produces each result. Change any of them and you can get significantly different outcomes.

There are basically two reasons why surface finish gages might not correlate. The first involves the gage settings. On most gages there are ways to select cutoff length, the number of cutoffs, the filter to be used, traverse length and speed, or even what type of probe should be used (skid or skidless). To correlate these, make a list of the switch and dial position settings of one gage—almost like a checklist—and compare it to the settings of the other gage. One-by-one, make sure all settings are the same. On newer gages you may be able to print out all the settings. Use this and go down the list to make sure they are all the same.

The second involves the condition and configuration of the gage. This includes the tip radius (there are a number of standard tip radii); the tip condition (is it worn or cracked?); whether or not the gage is calibrated; and its measuring force. Environmental conditions need to be looked at here as well: is there a source of vibration entering into the measurement? Run the measurement without moving the probe: do you see some results?

This could be the measurement of vibration coming through the base granite and into the gage.

Once you have checked the settings of the gage and its physical condition, there are a couple of "hidden" influences that may also affect the results from each gage. Newer gages have updated filtering capabilities compared to older versions. These filters can separate out high frequency noise from the measurement. Check to see if these, $\lambda c/\lambda s$ (lambda), filters are part of each measuring system. Finally, in the world of surface finish parameters, we know that different parameters sometimes have the same name. R_z, is a perfect example. This parameter has a number of different meanings (and algorithms) in different countries and standards. Always check to see that the parameters in each gage are using the same method of analysis.

Once there is confidence that each gage is set identically, there should be "fairly" good correlation between the systems. But, remember, there will always be some slight variation, especially when analyzing very complex surfaces.

■ ■ ■

Hardness Testing And Surface Variation

Hardness measurements are a topic we have not previously addressed. But when dimensional measurements can affect the hardness results, it's time we should think about it.

Although there are any number of ways to define and measure hardness—including some new methods using ultrasonics—the most common form of hardness testing is the static indentation as measured on several scales, including Brinell, Rockwell, Vickers and others. The basic idea of the static indentation test is that a set force is applied to an indenter in order to establish the resistance of the material to penetration. Should the material be hard, then a relatively shallow indentation will result. If the material is soft, then a larger or deeper indentation will occur.

The main difference between the several types of static indentation tests lies in the shape of the indenter. These can be round, pyramidal, or triangular, but the principle is the same: simply, a force is applied by means of the indenter and the size of the resulting mark is related to a hardness value.

There are many things that can influence hardness results aside from material considerations. These can include differences in indenter shape, the relative consistency of the force being applied, the velocity of the probe, and whether the operator is using the test equipment properly to assure that the force is being applied square to the surface.

Another big source of error with some hardness testing is the quality of the surface on the test sample. This is where surface finish inspection starts to play a role. When reviewing the literature of many hardness gages you will see a minimum surface finish requirement specified in order to assure proper hardness gage function. Surface finish of 80 microinches or better average roughness (R_a) are often required to assure proper hardness measurements.

Surface finish—also known as profile—is composed of two elements: waviness and roughness. Waviness, or longer-wavelength variation, is caused by macro-type influences, such as worn spindle bearings, or vibration from other equipment on the shop floor. Roughness is the short-wavelength pattern caused by tool marks from grinding, milling or other machining processes, and is influenced by the condition and quality of that tooling. As the indenter is apt to be small compared to the waviness component of the surface, it is this latter, short-wavelength roughness pattern that influences hardness values the most.

Average Roughness, or R_a, is the most widely used parameter for measuring surface finish. It is an arithmetic average of surface irregularities measured from a mean line that lies somewhere between the highest and lowest points on a given cut-off length. Today there are more than 100 additional means or parameters with which to calculate surface finish from the same measurement data, but R_a is the most common and is available through most entry level surface finish gages.

It is pretty easy to visualize what happens when the indenter of a hardness tester impacts on a smooth surface: the point—whether rounded or sharp—penetrates, and the area of contact increases geometrically until it reaches the depth associated with the hardness

Section C: Surface   81

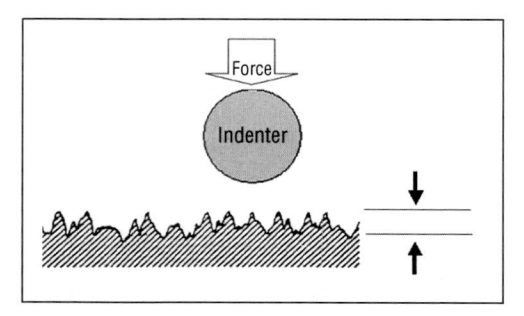

The indenter has to overcome the resistance of the peaks before it reaches the surface. This varying resistance will affect the hardness test.

of the material. However, when that surface is rough, a number of different forces come into play.

If the surface has a consistent pattern of roughness, then the indenter must work its way through a succession of more or less evenly spaced peaks as its force is dissipated. So besides the hardness in the part, the indenter is seeing increased resistance as the probe increases its area of contact with the material. On the other hand, if the surface has some big peaks in it, the indenter will hit these first, then work through lesser peaks before it gets to the "real" surface.

Either of these conditions—and any number of others—will add variability to the hardness measurement result.

How do you reduce the variability? Make sure the print for the part calls out a surface tolerance. Verify that the surface call-out is less than the requirements needed by the hardness tester. And then use a gage to verify that the surface is within spec prior to making the hardness test.

This process is no different from any other dimensional measurement. When checking gage blocks you need to be sure each block is clean, has no burrs and has reached the proper measuring temperature. With hardness, the measurement process spec should require the operator to verify that the surface is clean and has no burrs, and that a surface finish check be done to ensure it is within the tolerance of the part and of the hardness testing gage.

Once you're sure the surface finish is within limits, variation in your results won't make you nearly so tempted to test the hardness of your own head.

Section D
Electronics

Amplifiers: More Than Just Readout Devices

An electronic gaging amplifier is one of those devices whose full potential is rarely appreciated by its owner—sort of like Range Rovers that never leave the pavement. Gaging amplifiers are often used simply as replacements for dial indicators where a higher degree of resolution is required. This is to ignore numerous opportunities to make gaging more efficient and productive.

Some amplifiers, for example, incorporate dynamic measurement capabilities, including Minimum (Min), Maximum (Max), and Total Indicated Reading (TIR) functions. The amplifier "remembers" the highest and lowest points measured on a part, and displays either or both of them, or subtracts the Min from the Max to calculate TIR.

This is useful when gaging round parts in a V-block fixture, or measuring the height of a flat surface. The operator can quickly turn a shaft through a complete revolution, or move a flat part around under the gage head, without pausing to read the display. When manipulation of the workpiece is complete, the operator may select to display the maximum or minimum ID, OD, height, depth, or runout.

Other advanced functions can speed gaging setups. The "auto-zero" function is the electronic equivalent of the rotating bezel on mechanical dial indicators: the operator brings the gage head into rough contact with the master, and simply zeroes the amplifier, eliminating the need for ultra-careful positioning of the gage head. A "master deviation" function allows the addition of a fudge factor to the zero setting. Say your spec calls for a nominal dimension of 1.99980", but you've only got gage blocks handy for 2.00000". No problem. Simply set your zero at 2.00000", master the gage, program in a deviation of +.00020" to all measurements, and voila! Quick and easy mastering, without the hassle of post-measurement arithmetic.

A "preset value" allows switching between comparative and absolute measurements. In other words, instead of gaging deviation from nominal, the amplifier displays actual part dimensions. (In the above example, if a part is .00010 above nominal, the display will read 1.99990".)

Many amplifiers accept signals from two or more gages. This means that more than one part feature can be measured on a multi-gage fixture, by simply "toggling" between the inputs. Somewhat more sophisticated is the capability for differential measurements, in which the amplifier subtracts the reading of one gage head from the other: for example, you can derive straightness by calculating the difference in height of two co-linear points on a shaft.

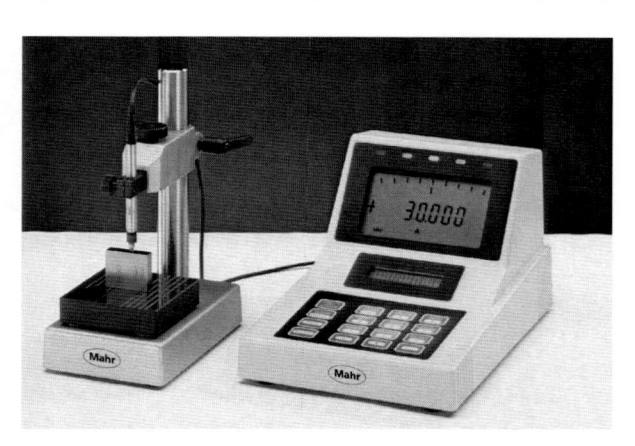

Amplifiers also allow the user to establish tolerance limits, and some incorporate green, amber, and red lights to indicate "in tolerance," "approaching limits," and "out of tolerance" conditions.

Alternately, the lights can indicate different part-size categories for match-gaging applications. Through digital output ports, the same electronics can be used to drive large accessory lights, enhancing parts sorting efficiency or bad-part identification in high-volume applications.

These digital output ports represent a great benefit of modern benchtop gaging amplifiers. Through them, gaging data can be used to control production machinery on an in-process basis, replacing expensive, dedicated closed-loop controllers at a fraction of the cost.

In one real, representative application, a gage head is positioned to measure a workpiece while it's still on a grinder. The user of this system assembled it using an off-the-shelf, hermetically sealed gage head, and a standard benchtop amplifier connected via the digital I/O ports to the grinder's computer-numeric controller. The grinder shifts to a shallower depth of cut when the gaged data approaches the specified dimension, and stops automatically when the spec is reached. As long as the system is calibrated adequately, no post-process gaging is required.

Besides these enhancements to the gaging process, the most important and widely used feature on modern amps is the RS-232 port for data collection. Now, through SPC, intelligent decisions can be made about the sample lot or the process. Amplifiers also provide analog output to drive strip chart recorders for continuous part measurement.

Not all gaging amplifiers incorporate all of the features listed here, although most modern amps incorporate some of them. When selecting a new amplifier, one can readily enough identify the product features needed to meet the requirements of the application. For those who are currently using amplifiers simply to take comparative measurements, it may be worthwhile to review the owner's manual, to look for built-in functions that can enhance your productivity.

■ ■ ■

It's An Analog World

In spite of what some Internet addicts may think, the world is analog, not digital. A simple example that proves this statement is what happens when we try to cross a busy intersection on foot. If the world were digital, we'd be limited to working with simple "on/off" information, indicating "car present/not-present." If we took several readings over time, we would be able to extrapolate, but not directly detect, a car's direction, speed, and acceleration. By the time we had done all that, that car would be long gone and we'd have to start collecting the next data set. But because the world is analog, a brief glance is all we need to detect presence, distance, direction, speed and acceleration. This enables us to react safely, either by staying put or crossing, choosing our rate of acceleration somewhere along a continuous but finite scale of values.

Analog gaging devices also contain more information than digital ones. Just watching the sweep of the needle across the dial of an analog amplifier, from "a little on the plus side" to "a little more on the plus side" may provide a machinist with all the information he needs to make the right decision to maintain control over his process—even if he doesn't actually read any numbers from the dial. So in spite of the benefits of digital instruments (more on this below), analog systems still have an important role to play.

Analog amplifiers excel in "dynamic" applications, where the gage head moves relative to the part (or vice versa). For example, when "exploring" parts for flatness using surface plate methods, the user slides the gage stand around on the plate, and quickly observes the amount of variation in the part. If the user had a digital amplifier, he would position the stand, wait for a moment to read the value on the display, reposition the stand, read the display a second time... and so on, until a sufficient number of data points had been collected.

The same principle applies to measuring out-of-roundness, in which the part is turned on a V-block beneath a stationary gage head. Using an analog amplifier, the user can directly observe the amount of variation, compare part size to the mastered dimension, and see whether the variation is all on the plus or minus side, or balanced around zero.

Machine tool setup is another valuable application for analog amps. For example, to ensure good centering when preparing to final bore a hole, a lever-type gage head is mounted on the machine spindle, with the contact against the inside of the bore. By turning the spindle back and forth by hand, and making cross-slide adjustments while watching the movement of the amplifier needle, the user can readily center the bore directly under the spindle. The same principle applies to positioning a fixture on a machine table by means of a reference "button" on the fixture.

Some analog amplifiers accept inputs from level-sensing devices, in addition to dimensional transducers. Electronic levels are of value when installing a new machine or truing up an existing one: the analog display enables the user to watch the effects of leveling adjustments in real time. If the amplifier has dual inputs, two levels can be used in tandem for "differential" measurements, to check parallelism or squareness between surfaces.

When selecting an amplifier for dynamic applications, look for adequate response speed to display change as soon as it occurs. Also, consider the needle's tendency to overshoot the measurement during rapid changes: some amps control this better than others.

Other important features include: switchable "normal/reverse" settings to make setup and interpretation easier; a dial with selectable range/resolution settings to accommodate a variety of tolerance specifications; and an analog output port for collecting data on a strip chart recorder.

None of this is to imply that analog amplifiers are always the best choice. Digital systems are superior for the purposes of data processing and output, and the generation of feedback for the control of CNC machines. Some digital amps incorporate "dynamic" features that automatically capture minimum or maximum readings, or calculate the difference between the two (i.e., TIR). Although the digital amp user can't see the sweep of a needle, he can still obtain "variable" information fairly readily.

In a nutshell, digital devices are generally preferable where:
- The measurement is static—neither gage nor part are moving.
- Output for data collection or analysis is desired.
- High resolution over long range is required.

Analog devices are preferred when:
- The measurement is "dynamic," involving a moving part or gage.
- You want to observe trends or rates of change, as in approach-to-size, leveling, positioning, flatness, and out-of-roundness measurements.
- High resolution over short range is required.

■ ■ ■

Three Heads Are Better Than One

Electronic gages are the instrument of choice for many demanding inspection applications. Digital electronic amplifiers offer high resolution, excellent stability, and the ability to output to data collectors, and can be integrated into feedback-controlled manufacturing systems. Furthermore, they can be programmed to capture minimum or maximum readings, to calculate TIR and average ("nominal") measurements, and to combine readings from more than one gage head, in addition to numerous other options and features that vary with make and model.

Users of electronic amplifiers can choose from a number of gage head types to generate the measurement signal. Cartridge, pantograph, and lever-type gage heads are the three most common, differing from each other mainly in the orientation of their sensitive contacts, and the mechanisms by which contact movement actuates the transducer. Other types of dimensional sensing devices that can be integrated into electronic amplifier-based systems include capacitance gages and laser devices, and these can be useful in applications that require non-contact sensing. But by far the greatest number of metalworking applications are satisfied with the three common "mechanical," or contact-type gage heads. Each of these has particular advantages for different applications.

The cartridge probe, or pencil-type gage head, is a compact cylindrical package, usually $^3/_8$" or 8 mm in diameter. Not coincidentally, these are the same diameters as dial indicator stems, and the cartridge probe was designed for direct replacement of indicators. Like dial indicators, the probe's spindle, or sensitive contact, has an axial motion. Readily incorporated into fixture gages and in-process gages, several cartridge probes can be positioned within close proximity, to measure closely spaced part features. For even tighter spacing requirements, some manufacturers offer special miniature probes, with diameters as small as 6 mm and lengths below 20 mm.

Numerous other options and variants are available to increase flexibility of application. Measurement ranges vary from as short as ±0.010" (±0.250 mm) to as long as ±0.100" (±2.5 mm), with linearity ranging from ±0.05% to ±0.5%. Longer ranges are available, but they are usually

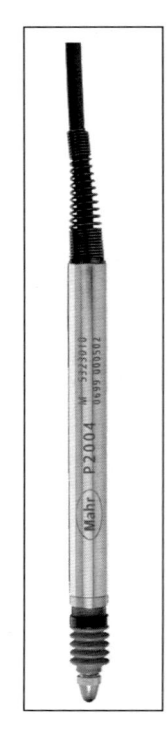

Section D: Electronics

not applied for tight-tolerance measurements. (Linearity is typically a trade-off against longer range.) The standard plain bushings that support the spindle tend to be quite durable, but in applications that subject the spindle to significant side loading, ball-bearing bushings can provide longer life cycles. The signal output cable is normally supplied straight and plastic-jacketed, but coiled cable is available for use on hand-held gages, and armored cable is available for harsh environments. Cable may exit the probe from the back of the cartridge (axially), or at a right angle—a small detail that occasionally makes mounting the probe much more convenient.

Most cartridge heads are splash proof, with a protective rubber boot surrounding the probe's stem. Hermetically sealed versions are also available, for use in extremely harsh environments: for example, for in-process gaging during a grinding operation.

Cartridge heads tend to have relatively heavy gaging pressure—about 3.5 oz. (99 g)—but here too, optional specifications are available from some manufacturers. Another handy option is a pneumatic retraction accessory, to minimize side loading on the spindle when inserting a workpiece into a gage fixture.

Pantograph, or reed-spring gage heads, are most often used in benchtop height comparators where both ruggedness and extremely high accuracy are required. The gage's contact is suspended by a pair of reed springs, which provide virtually force-free and friction-free measurement. (External springs or deadweights can be added if a specific gaging pressure is required.) Pantograph gage heads offer a measurement range of ±0.010" (±0.250 mm) with repeatability of <0.5 microinches (<0.01µm). They are more accepting of side loading than cartridge-type gage heads, and can be repaired more easily and economically if side-loading damage does occur.

Where the cartridge gage head replicates the action of the dial indicator, lever-type gage heads are functional replacements for test indicators. Electronic lever-type gage heads are typically used in connection with a height stand, often for surface-plate work. When mounted on a tiltable, extendable crossbar, they can be positioned with a great deal of latitude relative to the workpiece. A clutch on the swivel further assists in positioning convenience, allowing the contact to be repositioned by as much as 20° without moving the body of the gage head. Their ability to measure in both directions further enhances versatility. In contrast, cartridge and pantograph gage heads are uni-directional, and must be positioned perfectly in-line with the dimension being measured.

The extended, pivoting contact of the lever-type gage head provides good access to working surfaces that may be hard to reach with other contact styles. Contacts with special shapes, and diameters as small as 0.010" (0.250 mm) can be specified for use on really inaccessible workpiece surfaces. Repeatability can be as good as <4µ" (<0.1µm), and gaging pressure, at <0.14 oz. (<4 g), is light, making these heads well suited for high-resolution measurements on delicate surfaces or compressible materials.

All three types of electronic gage heads can be combined in a single fixture gage or application, and many amplifiers will accept all three interchangeably. All three can also be readily integrated with either digital or analog amplifiers. These features help make electronic gaging a very flexible approach to high-accuracy inspection.

■ ■ ■

Electronic Gaging Basics

Mechanical gages are familiar and economical. Air gages offer non-contact measurement and ease of use. But for the highest levels of accuracy and performance, it's hard to beat electronic gaging. No other method combines all this: extremely high resolution; relatively long range; adjustable magnification; programmability; digital output; and flexibility to move from job to job.

The "basic" electronic gage consists of three elements: a gage head; an amplifier; and a fixture or a stand to position the gage head relative to the workpiece. We'll touch on each separately.

The most common gage head type is the LVDT (linear variable differential transducer), an electromechanical device consisting of a primary coil, flanked by two secondary coils connected in series, all surrounding a movable magnetic core (the spindle), which provides a path for magnetic flux linking the coils. When the primary coil is energized by a sinusoidal signal from the amplifier, voltage of opposite polarity is induced in the secondary coils. The device's net output is the difference between the voltages of the two secondary coils, so when the core is centered, net output is zero.

The null or zero position is very stable, making LVDTs ideal for high repeatability comparative measurements. And because the LVDT works on an inductive principle, its resolution is, in theory, virtually infinite. In practice, it is limited by the amplifier's ability to amplify and display the results. Ranges vary from ±0.010" to ±0.100" (±0.250 mm to ±2.500 mm), with linearity from 0.5% to 0.05% over the nominal range.

Even standard-duty LVDTs are very rugged, and heavy-duty versions are capable of extended use in the harshest environments. Less than 3" (75 mm) long and about the diameter of a pencil, they can be laid out with great flexibility in fixture gages.

Gaging amplifiers are made in analog and digital versions. Analog amps are preferred where highest resolution is required (at short range); where multiple-range capability is desirable in a single task; where measurement involves watching trends (such as approach-to-size); or where part motion or exploration is required (e.g., measuring flatness over a large area). Digital amps are preferable where high resolution and relatively long range are required; where the measurement is static (no motion between part and gage); and where digital output is required for data collection or machine control. In general, analog amps tend to be used in machine setup and surface plate measurements, while digital amps are used in high-volume inspection applications.

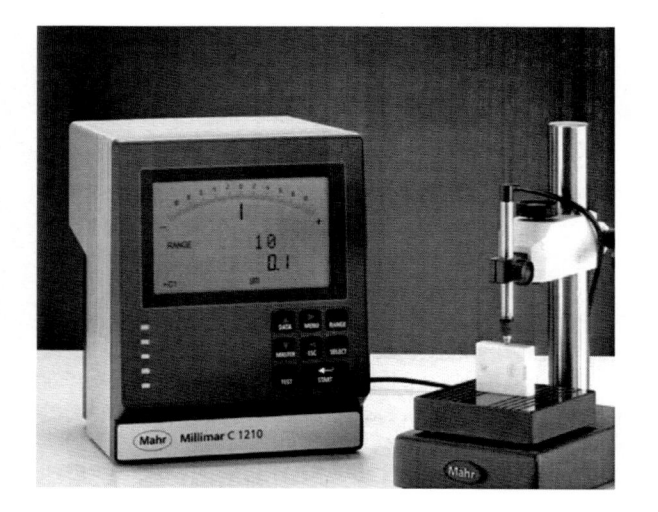

Many amps have two input channels, and the ability to combine signals from two transducers into one measurement is another benefit of electronic gaging. In "differential" mode, the amplifier is programmed to add or (more commonly) subtract one signal from the other. This gives the gage the flexibility to measure parts when mechanical references are difficult or impossible to establish, or when two variables may exist simultaneously: for example, on nominally round parts that are subject to dimensional variation and out-of-roundness. Depending on gage setup, the amplifier can be programmed to display either or both variables.

A typical digital amplifier might offer a choice of three measurement range/resolution combinations: ±0.100"/0.0001"; ±0.010"/10 microinches; and ±0.001"/1 microinch (±2 mm/0.001 mm; ±0.200 mm/0.0001 mm; and ±0.020 mm/0.00002 mm). It is important to remember that microinch/sub-micron resolution on the display does not necessarily mean that level of accuracy in practice. The accuracy of the fixture and the master, the geometric consistency of the workpiece, the stability of the gaging environment, and other conditions will influence gaging results. In practice, accuracy below 10 microinches (0.000025 mm) generally requires a laboratory environment.

In order to achieve the best accuracy under shop-floor conditions, it is essential that the gage headstand be as sturdy and stable as possible. Platen, post, arm, and mounting bracket must all be totally rigid when locked. For high-precision height gaging, the arm assembly is often equipped with a lead screw mechanism, because the extra-sturdy arm is too heavy to be conveniently raised or lowered manually.

Another great benefit of electronic gaging is the ease with which components can be swapped around for different jobs. An amplifier that is used with two LVDTs for differential measurements in a fixture gage today may be hooked up to a single lever-type gage head in a comparator stand for height gaging tomorrow. Likewise, a gage head may be used with analog amplifier one day, and with a digital amp the next. No matter what your dimensional gaging problem, an electronic gage can probably be configured to handle it without too much fuss.

Gaging By Computer

In spite of the proliferation of personal computers in almost every other industry area, PCs are still somewhat of a rarity in gaging applications. Computers are rarely necessities for standard dimensional measurements, although almost any application can be enhanced through the use of PC-based gaging software. And while the use of "gaging computers" cannot serve as a substitute for sound gaging practice, the potential benefits they offer are greater, and the barriers to entry lower than ever.

At about $5,000 for a complete hardware and software package, a gaging computer is often cost-justified strictly on the basis of its ability to display multiple dimensions simultaneously, compared to the cost of multiple amplifiers or column gages to provide the same display capabilities. The break-even usually occurs at just four or five simultaneous measurements; as the number of measurements increases, the PC becomes ever more economical.

For simultaneous measurements, multiple "light bars" can be arranged on the monitor, each bar growing or shrinking in height in response to the part dimension. The display can be programmed for "stoplight gaging," in which the bar changes color as the measurement shifts between in-tolerance (green), approaching limits (yellow), or out-of-tolerance (red). While measurements appear simultaneously in a numerical format beside each bar, a quick glance to verify that all the bars are green usually suffices to confirm the stability of a process.

Alternately, the monitor may display the part print, with measurements appearing call-out-fashion beside each feature. This view may help the operator understand more readily where process adjustments are required. If he wishes, he may "toggle" between the different displays.

By mathematically combining signals from multiple gage transducers, PCs can be used for complex measurements such as flatness, parallelism, squareness, taper, and distance between centers, or to perform simultaneous diameter and out-of-roundness checks on a single feature. PCs typically have 16 or 32 input/output ports (usually expandable), and support essentially all mathematical operations, allowing flexibility in the number and arrangement of checks. They also offer all the capabilities of digital gaging amplifiers, including auto-zeroing, adjustable resolution, reversible polarity (i.e., plus/minus direction), and dynamic functions like Min, Max and TIR. Most amplifiers, however, permit only two simultaneous inputs, and their range of mathematical functions is typically limited to +, -, *, and /.

PC programming and display options are highly customizable, with multiple options accessible through passwords. Operation can be made as simple or sophisticated as the application requires. For less-skilled users, the program may

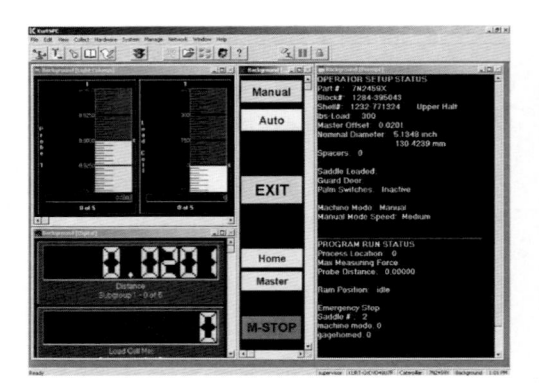

Section D: Electronics

boot directly to the required application, automatically capture measurements from the gage, and require no ongoing operator intervention other than loading the part on the gage.

At a higher level, software can lead the operator step by step through complex setup and measurement procedures, presenting instructions through a combination of text, graphics, and animation. Logic functions and event/action programming can help operators make decisions. For instance if the inspection task involves three hole diameters, the software may be programmed to give different instructions if one, two, or three of those holes are out of tolerance (e.g., "rework," "scrap," and "call manager," respectively). The same programming can be used to trigger automatic parts-sorting devices, or feed back to production machinery for closed-loop process control. In this type of application, the PC may replace multiple PLCs.

While there are numerous freestanding statistical process control packages, gaging computer software usually bundles these functions as well, easing the process of transferring data for the generation of X-bar and R charts, histograms, and all the other elements of SPC. Measurements can be automatically tagged with production data (such as machine number, operator, date and time, and events such as broken tools and coolant changes), and sorted and analyzed accordingly. SPC results can be displayed simultaneously with the results of an individual gaging trial, for a micro- and macro-view of the process at any moment. Finally, the ability to network PCs can eliminate the need for roving technicians with portable data loggers, and gives managers the ability to monitor all shop-floor processes from a central location.

Just a few years ago (when the world was still MS-DOS), PCs were intimidating to many users, and required substantial training to generate benefits in the hands of shop-floor users. That has changed, partly because the "average" machinist now does on-the-fly CNC programming, and partly because of the proliferation of Windows®. With their numerous cost-benefits and the elimination of the "fear factor," gaging computers should soon become a common productivity enhancer in machine shops.

Electronic Height Gages

In other columns, we've looked at "basic" comparative height gages, which are used for layout tasks and other surface plate measurements. These consist of a comparator stand, plus a test indicator or an electronic gage head and amplifier. Related to these are instruments known as Electronic Height Gages. These offer a high degree of flexibility and functionality, so that, in addition to lab-based work, they are useful as production gages. In quality departments, they are used for first part and incoming inspections, and layout work, while on the shop floor, machinists use them for checking features on one-off parts.

The key features of the electronic height gage are: a probe that senses when the part is touched; a glass or capacitance scale that tracks the probe's height; and a readout/control unit. Many also incorporate a motor drive to position the probe. There is a base and a body, to maintain the components in a stable, rigid relationship, and to accurately position the scale perpendicular to the surface plate on which the gage rests. And often, there is an internal pump that generates a thin cushion of air beneath the base, allowing the gage to be moved around easily on the surface plate.

Glass and capacitance scales have gotten so good that these gages are reliable enough for shop-floor use, and so accurate over a long range as to blur the lines between comparative and absolute gaging. Most height gages can measure in both modes, and even toggle between them on a single measurement. Resolution of .0001"/.001 mm, with accuracy of .0005"/.013 mm over a range of 24"/615 mm is common, while high-end instruments offer resolution down to 10μ"/.5μm and accuracy of .00012"/.0025 mm.

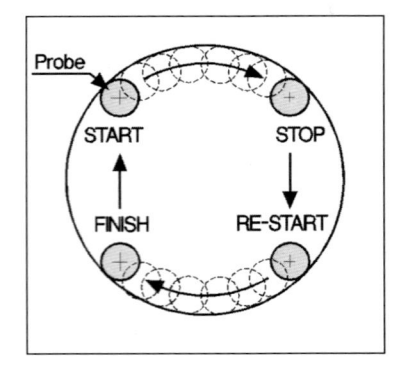

Two sensing technologies predominate: touch triggers and active probes. Both types can be set to trigger from both downward and upward touches. Once these points are collected, it's easy to calculate the difference between them, for either inside measurements (such as slot lengths and widths, and inside diameters) or outside measurements (such as ODs or thicknesses). One can also average the two readings to find hole centers or centerlines. From there, it's an easy step to calculate distances between centers.

The more common touch triggers send a signal to the scale only once per touch. Active probes, found on higher-end systems, constantly update their position, and record the position once they reach a stable reading on the part. In addition to single-point measurements, gages with active probes can be used for "dynamic" measurements, to explore a feature for straightness, flatness, MIN, MAX, or TIR.

Active probes have the potential to generate more accurate diameter measurements, because the user can tram the gage perpendicularly to the feature's axis, to capture the highest and lowest points on the top and bottom surfaces. (See figure.) To correctly measure a diameter with a touch trigger, a special contact is used, which is designed to seek the low or high point of the diameter.

Even the relatively simple control units associated with touch triggers tend to be highly capable. These are usually programmable for multiple measurement routines, can accept presets, and calculate widths, thicknesses, and distances between centers.

More powerful controllers, which usually accompany active probes, are required for dynamic measurements. These data processors are capable of generating SPC reports, and turning the single-axis height gage into a virtual two-dimension measuring machine. One can measure bolthole patterns and similar 2D relationships, by measuring the height of the holes, then reorienting the part 90° and remeasuring the hole heights again. The controller includes a one-button "90° flip" function to calculate results as X-Y coordinates (e.g., a hole center is 6.000" in from one edge, and 2.000" from an adjacent edge) or as polar coordinates (e.g., a hole center is 16.342° from a reference point, on a hypotenuse of 4.500").

The gage must be zeroed before measuring parts. This is usually done by touching the probe to the reference surface—usually a surface plate. Gages can also be referenced against a gageblock, or against a datum on the workpiece itself.

Before measuring inside or outside dimensions, the diameter of the ball end of the probe must be compensated for. This involves touching the probe to the top and bottom of a special reference artifact. The controller calculates the diameter as the difference between the measured reading and the known distance between the two reference surfaces.

While general-purpose measurement devices like electronic height gages can't compete with some types of comparative gaging for measurements requiring very high resolutions or throughput, they are ideal for most surface plate layout work, and for inspection of parts produced in small quantities.

Section E
Calibration

Calibrating Gages: Your Place Or Mine?

All gaging equipment must be calibrated periodically to ensure that it's capable of performing the job for which it's intended: i.e., measuring parts accurately. This has always been necessary for the purpose of maintaining quality, but there are now additional, external reasons to establish and maintain a regular program of gage calibration: customers' requirements. More and more OEMs demand that suppliers document their quality efforts from start to finish. ISO 9000 is one more manifestation of this trend, and it is forcing companies to examine their calibration programs, identify their weaknesses, and improve them wherever possible.

Some large companies with thousands of gages can cost-justify hiring or training specialists in gage calibration methods and supplying them with equipment and resources to perform virtually all calibration duties in-house. For most machine shops, however, the economical approach is to hire a calibration service.

ISO 9002, which applies to all manufacturing operations, requires suppliers to calibrate "all inspection, measuring and test equipment and devices that can affect product quality at prescribed intervals, or prior to use, against certified equipment having a known valid relationship to nationally recognized standards—where no such standards exist, the basis used for calibration shall be documented." (ISO9002.4.10.b) Let's elaborate on some of these points:

"(P)rescribed intervals" usually translates into a minimum of once per year. Where annual calibration is inadequate to ensure accuracy, a shorter interval must be established.

"(C)ertified equipment having a known, valid relationship" means that the calibration house must have its own equipment certified. In the U.S., "nationally recognized standards" implies the National Institute of Standards and Testing (NIST), although other standards, such as DIN, may be used to satisfy overseas customers. "(W)here no such standards exist," usually refers to highly specific industries or products, where the manufacturer must develop his own standards and test methods—(say, a foam pad of known density, used to master a chocolate-pudding-consistency gage). Calibration houses issue a certificate of calibration for every gage tested. These certificates are essential for users to document their calibration programs. At minimum, they must include:

- The serial number and description of the gage tested.
- The serial number of the gage(s) used to perform the testing.
- The level of uncertainty of the calibration—in other words, the tolerances of the data.
- A statement of traceability to NIST (or other standard).
- A serial number identifying the NIST test upon which the calibration house's own standard is based.
- Reference temperature under which the calibration was performed.
- Name of the customer; name and address of calibration service.
- Date of calibration and signature of the technician.
- Test results: e. g., error in the gage, measured at appropriate intervals across its entire range.
- If the gage is adjusted subsequent to testing, it must be recalibrated, with results as above.

Some providers automatically remind their clients which gages need to be calibrated, and when. Most gages can simply be boxed and shipped to the calibration house, although in the case of large, elaborate gages (e.g., circular geometry gages, CMMs) the mountain must come to Mohammed. The calibration service will come prepared with NIST-traceable gage blocks, precision balls, a thermometer, and any other standards needed to perform the job.

How can a machine shop without expertise in calibration intelligently select a provider? Naturally, cost and turnaround time are important, but don't sacrifice quality for convenience. Above all, ISO 9000 requires that consistent procedures be applied, and any professional calibration house should be able to document its methods in a procedures manual. Ask to see it, and if it's unavailable, look elsewhere.

Surprisingly, there are no certification standards for calibration labs themselves, so a supplier's reputation is important. Don't be afraid to ask questions—lots of them. What are his areas of expertise? How are his technicians trained, and what is their level of experience? What test equipment is used, and to what standards can test methods be certified (e.g., MIL, GGG, ANSI)? What quality control methods are in place? What is the physical design of the facility, what are the control tolerances on temperature and humidity, and how is the equipment protected from the effects of outside vibration? The figure shows an example of state-of-the-art isolation from dynamic forces: how does the facility under consideration compare? A visit may be well worthwhile.

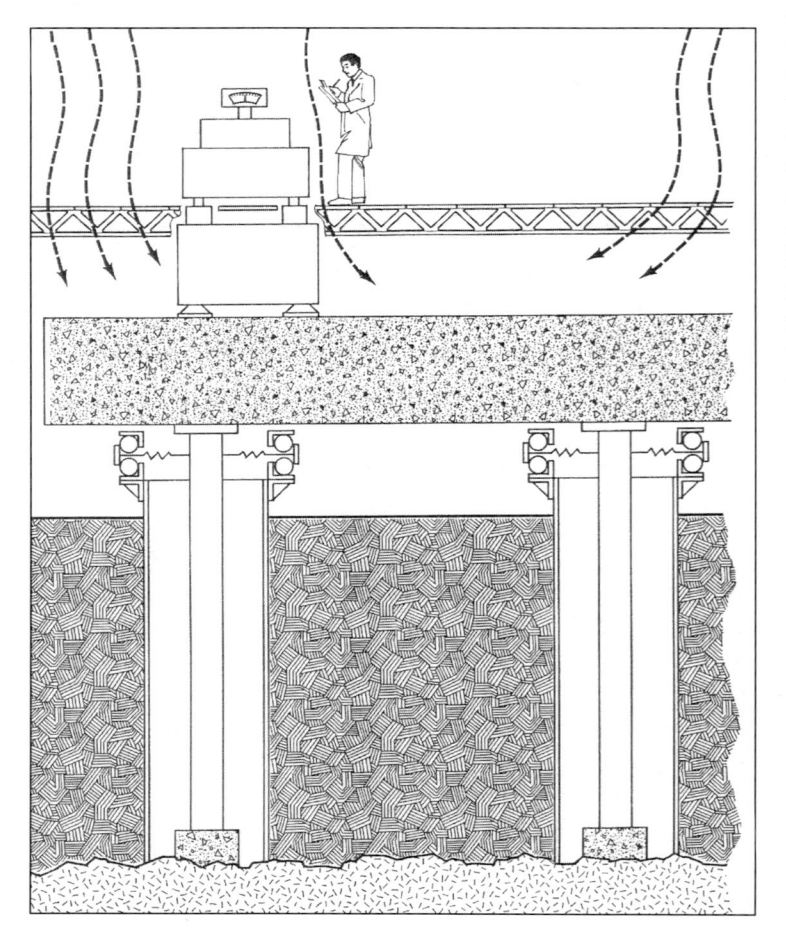

Gages in this calibration center rest on a slab that is isolated from the building itself. The slab is supported on pilings down to bedrock and the pilings are isolated in caissons.

Mastering For IDs And ODs

Once upon a time, an overly enthusiastic QC manager appealed to me, confused and dissatisfied. Here he was, spending good money to purchase very high quality masters, but his inspection process was no better than before. What was worse, his masters went out of calibration rapidly, pushing his costs even higher. The problem was that he was buying more accuracy than he could use.

Choosing the right tool for the job applies to mastering, just as it applies to every other area of gaging. While it may be possible to master a gage using a variety of standards, the best master for a job strikes a balance between accuracy, economy, durability, and ease of use.

Gage blocks are "primary standards," directly traceable to an "absolute" standard maintained by NIST, DIN, or ISO. Masters are "secondary" standards, because their sizes are established by reference to primary standards. While masters typically have a higher level of uncertainty than gage blocks, they are often the appropriate choice for production gaging. Gage blocks, after all, are square, while masters are typically round. If the parts being measured are round, and the gage is designed to measure round parts, the use of a round master will help avoid certain sources of geometry error.

A master ring or ring gage is basically a bore of a known dimension. The same device can often be used as a setting master for variable inside-diameter gages (such as bore gages, air tooling, and mechanical plug gages), for go/no-go mastering of fixed ID gages (such as a fixed plug gage), and for go/no-go OD inspection of male cylindrical workpieces.

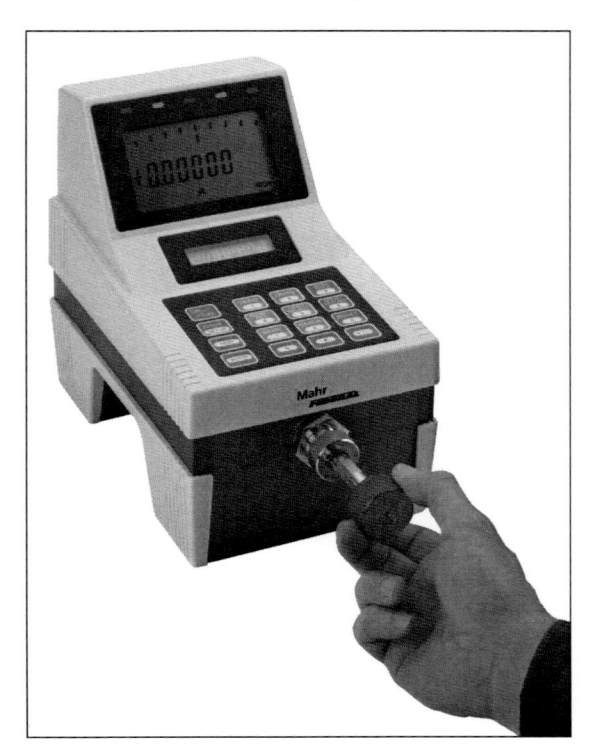

Ring gages are made from steel, chromed steel for durability and corrosion resistance, or tungsten carbide for extreme wear resistance. They are classed by level of accuracy, with XXX indicating the tightest tolerances, XX, X, and Y being intermediate grades (in descending order), and Z being the lowest level of accuracy. Class tolerances vary by size: larger sizes have higher levels of uncertainty. Tolerances may be bilateral (i.e., evenly split between plus and minus around the nominal dimension), for use in setting variable gages, or unilateral for use as go/no-go gages. For rings, "go" is minus (-); for plugs, "go" is plus (+). Go/no-go gages may often be identified by a groove or ring on their knurled outside diameters.

Plug gages, for go/no-go measurements of part IDs, or for mastering ID gages, are also available in different materials and classes. Plug gages may be reversible or double ended, with a "go" end signified by a green stripe, and a "no go" end signified by a red stripe. Usually available only in sizes up to about 0.76", reversible plug gages can be disassembled to replace a worn end.

Plug gages are often identified by the names of their handle or mounting designs. Taperlock plug gages usually range from 0.059" to 1.510", and have a handle on only one end. Tri-lock designs, also called discs, range from 1.510" to 8.010", and have handles on both ends of the mastering surface. Annular designs, for sizes from 8.010" to 12.010", are like wagon wheels, with handles for axles.

Specialty masters are available for a range of applications and odd shapes, including slots, splines, and tapers. Tool holder taper geometry is of increasing importance in precision machining, and manufacturers have begun to pay closer attention to taper quality. Taper plug gages can provide an indication of whether an ID taper is too steep or too shallow, or if the bore entry diameter is within tolerances. Inside and outside taper masters are also frequently used for setting taper air gaging. Such special-purpose masters make mastering and measuring quicker and easier, and usually cost more than standard gages.

In general, one should choose a master whose tolerance is 10 percent of the precision of the gage, while the gage's precision and repeatability should be 10 percent of the part tolerance. For example, if part tolerance is 0.001", gage precision should be 0.0001", and the master's tolerance should be 0.000010". It's usually not worthwhile to buy more accuracy than this "ten to one" rule: it costs more, it doesn't improve the accuracy of the gage, and the master will lose calibration faster. On the other hand, when manufacturing to extremely tight tolerances, a ratio of 4:1 or even 3:1 between gage and standard might have to be accepted.

Finally, here are some general guidelines for the care and feeding of masters: store them in a secure place; use a wax- or oil-based sealant to protect against corrosion; handle carefully—don't force or jam them onto the part; don't try to modify them; and when shipping for calibration, take steps to protect masters against damage and corrosion.

Control Thy Gages

The number of gages and micrometers that are in use, but actually incapable of doing the jobs to which they are assigned, is alarming. Too many machine shops make assertions of accuracy for a part or a process, based on gages that are scratched, sticking, or in some other obvious or hidden way incapable of taking good measurements. And when asked to document that assertion, these shops rely upon a dog-eared certificate of calibration that's years old.

While it may be alarming, the fact that inaccurate gages remain in use is not surprising. After all, there's simply no such thing as a "perfect" gage: even the best-engineered and well-maintained instrument has some degree of uncertainty. Every time the gage is used, components are subjected to some infinitesimal amount of wear. At what point does inaccuracy cross the line between acceptable and unacceptable?

Depending on many variables of design and usage, some gages retain accuracy for years, while others require refurbishing every few months. Some may be chugging along just fine, when an accident puts them suddenly out of kilter. But eventually, every gage loses accuracy.

Gage control—a system of record keeping used to track the use and condition of every gage in the shop—performs several important functions. Tracking when, where, and how each gage is used makes predictive maintenance possible, thus reducing scrap and rework. It's an important loss-prevention tool, helping to maintain your investment in valuable equipment. In the event that a number of bad parts slip through some level of quality

---

Mahr Federal Inc.
Gage Detail Report

9/12/97
Page 1 of 1
Gage ID: EMD-300P-1 **Gage SN:** FPC-0637
Supplier Code: FPC
   Cost: $650.00
Asset No: 012237
Purchase Date: 8/9/94
Model No: EMD-300P-1
Cal Hours: a
Manufacturer: Federal Products
Next Due Date: 12/1/97
Owner: Federal Products
Last Cal Date: 12/1/96
Description: Snap Gage
Status: 1
Active
      Type: Snap Gage w/ Maxum **Unit of_Meas:** inches
User Defined:
   Drawing No: EMD-300P-1 **Drawing Date:** 8/9/91
   Ref Standard: No Calibrated By: R. Smith
Usage:
Change Level: Rev. C
      RR Freq: 12
RR Freq Units: MONTHS
      Change Date: 5/17/95 **Storage Location:** Tool Crib 1
Current Location: Dept. 52
      Service Date: 8/9/97
Last RR: 12/1/96 **Next RR:** 12/1/97
Retirement Date: 8/9/00
RR Result: .0001
      Uncert: 0.00004
      Calibrator: CAL-0002 **Cal Frequency: 12**
NIST No.:
Notes:
Cal Frequency UOM: MONTHS
            Resolution: .00005"
   + Tolerance: +.0025
   - Tolerance: -.0025

control, it often permits analysis to determine how and when the problem occurred. Furthermore, it's essential to most relevant quality documentation programs, including ISO/QS-9000. (Although neither of these standards explicitly mentions how to control gages, it would be virtually impossible for a machine shop to demonstrate the required control over production without it.)

Although the gage control process is well defined by numerous company in-house and international standards, every shop must tailor the process somewhat to meet specific requirements. In large plants, it is often handled by the Inspection Department, which may establish a dedicated staff, facilities for gage calibration and inspection, and a gage storage crib. In small shops, the responsibility may be assigned to production or materiel managers, or the chief inspector. In any case, those responsible usually maintain daily contact with Process Engineering and Production, to define gaging requirements, establish budgets, maintain inventory, and calculate depreciation and obsolescence.

As a starting point, every gage and instrument should be assigned a unique serial number. The numbering system may be very simple, or particular digits in the serial number may be designated to reveal specific information about the gage. The control record should also include the date of purchase, the name of the supplier, and a description of the gage type, including the manufacturer's model number. If the gage was custom-built, the record should reference the engineering file.

The record should also contain answers to these questions:
- Where is the gage right now?
- When, where, and to whom was it issued?
- How long has it been on the job?
- How is it being used; on what product, and how often?
- What was its condition when issued?
- When was it last calibrated? How accurate is it?
- When is it scheduled to be recalibrated?
- What is its GR&R on a particular process?
- Has uncertainty been established?

Manual record keeping has largely given way to PC-based database software and specialized gage control programs, which have greatly increased the ease with which extensive records may be maintained and accessed. Among the many functions offered by commercially available programs, some automatically recall gages that are due for recalibration.

Historically, trade workers were often required to bring the tools needed to their jobs; this included machinists who were expected to provide their own gages. In many shops and plants, this practice still exists in modified form: the shop might provide gaging for inspection, but require machinists to provide gages for setups.

That practice is no longer viable. Manufacturers must be able to document procedures taken to assure quality. Unless 100 percent inspection is employed, this includes being able to demonstrate that setups were performed to a known degree of accuracy. This, in turn, requires gages whose accuracy is known. And that can only be done if the shop maintains control over all the gages in use.

A substantial investment is required to establish and maintain a gage control program. But through improved product quality, reduction of scrap and rework, loss prevention, tighter process control, and lower assembly costs, gage control almost always pays for itself over the long run.

Gaging And Mastering Uncertainty

When measuring parts to tolerances of a thousandth of an inch, we can usually be certain that our measurements are accurate to within a "tenth," as long as we follow standard gaging practice: i.e., master the gage frequently, maintain the gage in good working order, keep things clean, have the master recalibrated periodically, etc. But certainty becomes elusive at the microinch level. State-of-the-art machining practice is only just capable of producing gage standards and gage blocks to the required degrees of accuracy. However, their dimensions, as well as those of the workpieces, change readily with changes in temperature, the accumulation of infinitesimal amounts of dust, and minute variations in gaging practice.

Uncertainty can't be entirely eliminated, but manufacturers can successfully perform millionth measurements by relying upon relevant industry standards, which define how much uncertainty is permissible, and where. Particularly under ISO 9000, manufacturers must be able to document their use of reliable standards as the basis of their QA/QC efforts. But in all cases, uncertainty must be minimized, and one of the critical places to look for it is in mastering.

Gage blocks and masters have tolerances of dimension, surface roughness, and geometry: in other words, the masters themselves have inherent uncertainty. When gage blocks are wrung together, stacking error is introduced, combining all these sources of error with the added uncertainty that two or more wrings with the same blocks may produce different results. Gage blocks and masters are also subject to wear, which becomes significant rapidly at microinch tolerances.

Under the old "ten to one" rule, if you're measuring parts to 30 millionths, you want gage repeatability of 3 millionths, and a master that's good to 0.3 millionths. No one makes gage blocks to that level of accuracy, so we have to compromise and accept rules of five to one, or even less. That may be the best we can do.

Gage blocks are a "primary" standard: that is, they are documented and traceable back to an official, absolute standard—in the US, to the National Institute of Standards and Technology (NIST, formerly the National Bureau of Standards). Documentation makes it possible to determine the level of accuracy in a given gage block. Master rings and discs, in contrast, are generally considered to be secondary standards, because their size is established by reference to gage blocks. Traceability is thus one step further removed, which implies a greater level of uncertainty.

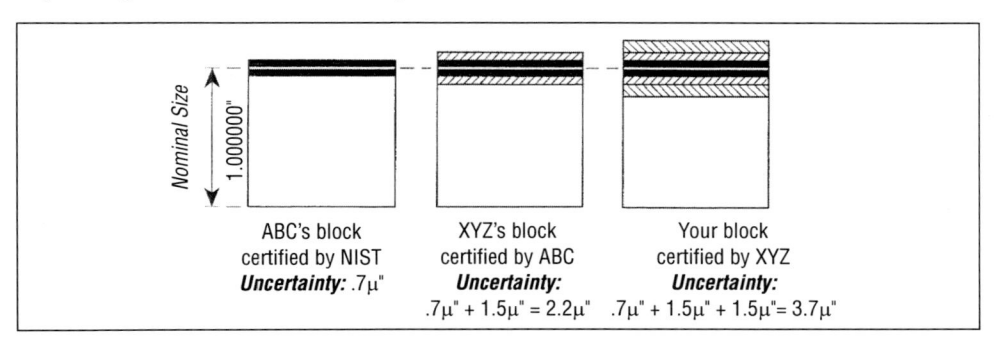

ABC's block certified by NIST
Uncertainty: $.7\mu$"

XYZ's block certified by ABC
Uncertainty: $.7\mu$" + 1.5μ" = 2.2μ"

Your block certified by XYZ
Uncertainty: $.7\mu$" + 1.5μ" + 1.5μ" = 3.7μ"

To document and minimize the level of uncertainty, gage blocks should ideally be sent to NIST for recertification. This way, you'll be mastering your gage at a single remove from the absolute standard: you can't get any closer than that. However, this may be impractical for a number of reasons, and commercial calibration houses may be able to provide faster service.

If you use a commercial service, it is important to choose one that sends its own primary blocks to NIST for calibration, to avoid adding unnecessary levels of uncertainty. Consider the following scenario:

You send your gage blocks to XYZ Accuracy Inc., but XYZ has its own blocks certified by ABC House o' Blocks. ABC sends its primary blocks to NIST for certification. Your blocks end up certified at three removes from NIST, with contributions of the following sources of uncertainty.

NIST uncertainty: 0.7μ"
ABC uncertainty: 1.5μ"
XYZ uncertainty: 1.5μ"
Total = 3.7μ"

While uncertainty isn't necessarily cumulative, it's easy to see how levels of uncertainty that may be insignificant for tolerances of .001" or .0001" can become critical when you're trying to measure to 10μ".

All this concern with mastering, calibration, and external standards is not an intellectual exercise of interest only to a chosen few: Any manufacturer hoping to meet microinch tolerances, obtain ISO 9000 certification, or satisfy many other industry standards, may be required to reference its measurement methods to officially recognized physical standards. Adequate traceability is an important issue, but one must be equally concerned with how many steps intervene between your own gage blocks and the official physical standard.

What's So Accurate About Uncertainty?

In talking about the sources of error that can be found in the setup of a form testing system, we frequently use the terms "accuracy" and "uncertainty." While these are often used interchangeably, every now and then, it's good to recall the precise meaning of these two words and how they affect the quality of measurements.

Accuracy of measurement is a phase that has been around for a long time. The internationally agreed definition for it says accuracy is, "the closeness of the agreement between the result of a measurement and the true value of the measurement" (for example a distance). In the official definition, it also notes that, "accuracy is a qualitative concept." This means it can be described as "high" or "low," for example, but should not be used quantitatively.

Often though, it is used quantitatively, as if the definition read, "the difference between the measured value and the true value." This leads to statements like, "accurate to ± X." The problem with this unofficial definition is that it assumes the true value can be defined and known perfectly. However, even in the deepest, darkest laboratories, backed by some of the most powerful countries in the world, perfect values cannot be realized. It's not a question of money or technology, it's just physically impossible to define or make a perfect measurement.

That's why metrologists use the term "uncertainty." The concept of uncertainty accepts that no measurement can be perfect, and is therefore defined as a "parameter, associated with the result of a measurement, that characterizes the dispersion of values that could reasonably be attributed to the measurand." It is, therefore, a range of values in which the value is estimated to lie. It does not attempt to define or rely upon a unique true value, but rather statistically estimates a range that value falls within.

Basically, what this comes down to is that using the term "accuracy" for the quantitative characterization of a measuring instrument is not compatible with the official meaning. In short, the way we use the word accuracy is technically improper and significantly different from the proper metrology term of uncertainty.

So what's better: to be proper, or to get the idea across?

In most situations, the difference really does not matter. It feels much better to say "the gage is accurate to…," rather than, "the gage is uncertain by….". This is probably because of the nature of the words. It sounds more impressive to say something is accurate than to say it is uncertain, and this is probably why we marketing types like it: we always try to sell the positive (although, we should perhaps use the word "certainty").

But, no matter what you call it, it is important that quality managers understand the concept of uncertainty and how it relates to gaging performance. Over the past five to ten years, there has been a lot of work done to develop ways of quantifying the performance of gaging equipment. This can be a complex task, even for the simplest gage, because it's not only the gage that influences the measurement; it's the standard, the workpiece, the people and the environment. All of these are subsets that potentially add some source of error to the measurement. So if you are trying to persuade others that your measurement result is a good one, you need to use all means available to express the concept of uncertainty.

To start—since the marketing people are the ones writing the specifications, you need to determine whether a gage's "accuracy" is likely to approximate the numbers given by the manufacturer. This is a real question and specifications should be interpreted with some consideration and caution. If the manufacturer has one concept of where and how the gage will be used and the purchaser has another, there will likely be a problem.

Bear in mind that the manufacturer's specification is often determined by measuring the most beautiful part, under ideal conditions, using the most advantageous part of the gage. You may be measuring rings that are rough, out of round, and tapered. So you need to develop a clear understanding of what "certainty" of performance you can expect under your specific conditions. Questions to ask include: what factors affect performance; how often does the gage need to be calibrated to maintain its performance; what are the required environmental conditions needed for the gage to perform to specification; do operators need special training; and in what condition must the parts be in order to make accurate measurements.

GR&R Measures More Than Just The Gage

A few weeks ago, a well-respected engine manufacturer approached me with a problem.

He was unable to pass a Gage Repeatability and Reproducibility (GR&R) study. The odd thing was that he's been using the same gaging method successfully for over 40 years.

GR&R is a way to assess the reliability of your gaging results. A GR&R study involves taking a few gage operators, and having each of them measure a small number of parts, several times each. The results are compiled, and (after some mildly confusing arithmetic) reduced to a single number that indicates the total expected spread of measurements for a single part, for all trials, by all operators. The number is presented as a percentage: a GR&R of 30 percent means that all the results fall within a range equal to 30 percent of the allowable part tolerance. (This is slightly simplified, but close enough for our discussion.)

In the case of the engine manufacturer, his target was a 10 percent GR&R on a part with a total tolerance of .001" (±.0005"). In other words, all the measurements for a given workpiece should fall within a range of .0001".

The manufacturer was using a hand-held snap gage to measure the part. Mounted on the gage was a dial indicator with a resolution (i.e., grad size) of .0001". Everything seemed to be in order. He was following the old gage maker's rule of thumb that states that you should have a 10:1 ratio between part tolerance and gage accuracy. He had been successfully measuring the part for decades using the same type of gage. The part hadn't changed. The tolerance hadn't changed. And yet, he was achieving GR&R results of 30-35 percent—not even close to the target.

He had failed to appreciate that something had indeed changed: his gaging requirement. Where previously a part would "pass" as long as it fell within a tolerance range .001" broad, GR&R now required that his gaging method meet a requirement much more demanding.

The problem wasn't his snap gage, which was in good condition, with a repeatability of 20 microinches. The problem was much simpler: he had the wrong dial indicator on the gage. With a resolution of .0001", the indicator itself ate up the entire allowance for variation under the GR&R study. And that left no room for the inevitable variation from other sources.

Remember the acronym "SWIPE"? There are five major factors that influence gaging results: Standard, Workpiece, Instrument, Personnel, and Environment. Each of these introduces a certain amount of variation to a measurement. Is the standard (the master) absolutely accurate? How about the workpiece's geometry? If it's out of round, it will generate different results every time you put it on the gage. The gage operators will introduce a certain amount of observational error, plus variability due to differences in gaging practice. Are you paying attention to the environmental factors that can influence a measurement: temperature, dirt, vibration, etc.? And, of course, there's the instrument—the gage itself—which could have stiction, wobbles, a misaligned holding fixture,

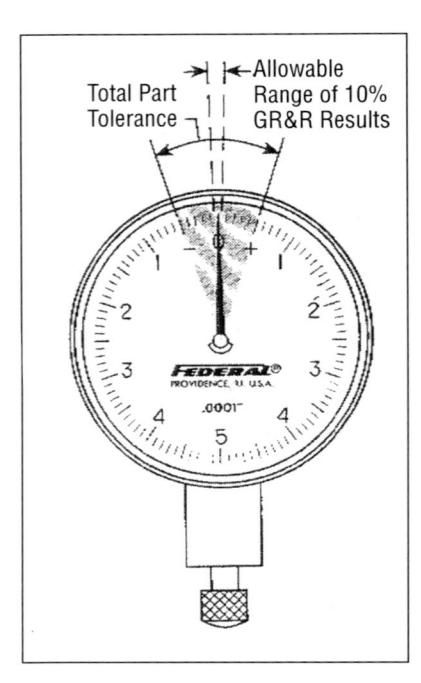

The gage in Fig. 1 could not pass a 10% GR&R study for a 0.001" part tolerance.

Section E: Calibration 105

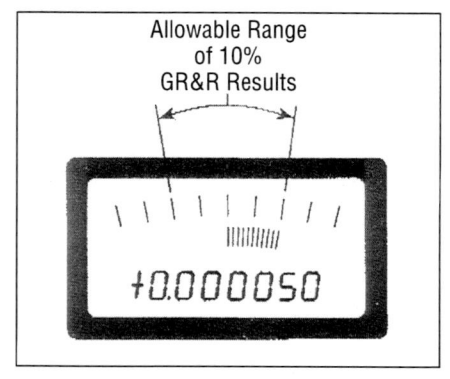

Fig. 2 shows one that could.

or even, just possibly, a wrongly specified dial indicator.

GR&R doesn't measure the gage in isolation: it measures the entire gaging process, with all of its influences and variables. If you want to achieve GR&R of 10 percent, then you'll have to be able to read the results to a considerably higher degree of resolution than 10 percent of the required tolerance. The old 10:1 rule is a general guide for a minimum level of accuracy—not an inflexible dictum for every application.

We replaced the dial indicator on the gage with an electronic probe capable of resolving to 50 microinches. This tightened up the margin for error imposed by the gage, and allowed room for other variables. The manufacturer was then able to meet the 10 percent GR&R requirement—without changing his manufacturing process, his gaging methods, or his gage.

If you fail a GR&R study, don't shoot the gage. You can't expect it to correct for errors from other sources. In fact, the moral of the story extends beyond the confines of GR&R. Any time you're assessing a gaging program or trying to determine your gage requirements, remember that the instrument is just one-fifth of the equation.

■ ■ ■

What's Wrong With This Picture?
On The Care And Feeding Of Master Rings And Other Metrology Artifacts

Quality Assurance can only be as good as the measuring tools it relies on. It should be obvious that if you spend tens of thousands of dollars on a measuring machine, you need to protect this investment with routine maintenance and calibration. The same is true for hand tools and gages, which are the nervous system of a manufacturing operation's quality system.

So dial indicators, hand gages and their masters need to be regularly calibrated. Checking these tools against recognized standards assures their reliable performance and provides for traceability when nonconformity does rear its ugly head in the manufacturing process. As soon as the manufacturing team buys into this concept and a program of regular calibration has become a way of life, your company will have taken a big step forward on the road to cost reduction and profit enhancement. That's the big picture.

If your quality assurance program is working well, it means everyone is taking care of the little details that are ultimately so important.

Newly calibrated gages, etc., are generally packaged and transported back to the floor with great care. That's a 'no brainer.' But what about the tools and artifacts that are being sent back to the calibration room to be checked again? It's very important to remember that even though these gages are out of service, they are still precision measurement devices. As such, they need to be handled accordingly.

Very often we will see gaging come back for re-certification in the condition pictured. Or, even worse, they will be all thrown into a box with nothing to prevent them from banging together.

Under a microscope, one good scratch on an XX master ring can look like the Grand Canyon, ruining an otherwise good master. And we discover these Grand Canyons with alarming frequency at our Precision Measurement Center where thousands of master rings and discs are measured in the course of a year. Many of these scratches result from the sort of treatment that the rings in the photo are subjected to.

If you don't think those rings are being abused, look again. For the most part this packaging was carefully applied, but notice the wire on the tags. Now, in some cases there may be some plausible excuse for the marking of the rings this way. Maybe they are badly worn and are being sent back to be lapped and chromed back up to specification. However, even though the wire is soft, it still will mark and potentially scratch the part. Therefore, this type of packaging is never recommended.

A natural alternative is to attach an identifying tag with string. But don't do it! String tends to absorb rust-causing moisture. Stamping them with tool numbers is not the answer either. The stresses created in the metal can sometimes ruin a master.

However, don't give up; there are ways to mark the masters without risking damage. Some acceptable ways of identifying a master for shipment and inspection are to mark them with paint or a permanent marking pen. Or include a sheet of paper that identifies the ring by its etched dimensions on the side. Even a sticky tag is a good temporary method of getting the ring identified until it reaches the source of recalibration.

Of the rings and discs that pass through our measurement center that have been sent in for annual size certification, surprising numbers have had to be reworked or even scrapped because of improper packaging. A ring needs to be sufficiently protected whether it is traveling across the shop or across the country. It should be protected with an oil and plastic dip, individually wrapped and sturdily packaged.

The safest policy is to have one or more individuals trained in the proper handling, packaging and transporting of hand tools and masters for recalibration. This is a simple little detail that can pay for itself many times over during the course of a year. Most gaging equipment suppliers would be happy to provide you with the guidelines you need to bring this little picture into sharp focus.

■ ■ ■

With Master Rings And Engagements, Size Matters

A master ring, or ring gage, is basically a bore of known dimension. The precision hole is often used as a setting master for variable inside-diameter gages (such as bore gages, air tooling and mechanical plug gages), for go/no-go mastering of fixed ID gages, and for go/no-go OD inspection of male cylindrical work pieces. Ring gages are made from steel, chromed steel for durability and corrosion resistance, or tungsten carbide for extreme wear resistance.

They are often classed by level of accuracy, with XXX indicating the tightest tolerances; XX, X and Y being intermediate grades; and Z being the lowest. Class tolerances vary by size. Larger sizes have more open tolerances since they are harder to manufacture. Tolerances may be bilateral for use in setting variable gages, or unilateral for use as go/no-go gages. For rings, "go" is minus; for plugs, it's plus. Go/no-go gages may often be identified by a groove or ring on their knurled outside diameters.

Example: for a 0.820" master ring the following tolerances would apply:
Class XXX=0.00001"
Class XX=0.00002"
Class X=0.00004"
Class Y=0.00007"
Class Z=0.00010"

Of course, the better the class the more you have to pay. If you want to stay at a 5-star hotel or get the highest grade for your engagement ring, be ready to pay for it. It's the same with master rings. The XXX ring is manufactured to a tighter tolerance, and there is cost involved with this. It may take longer to manufacture, take the skill of a higher paid technician, or if something goes wrong, it may have to be remanufactured and take longer to get.

Typically, the rule of thumb for selecting a master had been to choose one whose tolerance is 10 percent of the part tolerance. This, combined with the gage's performance, should provide adequate assurance of a good measurement process. It's usually not worthwhile to buy more accuracy than this "ten to one" rule: it costs more, it doesn't improve the accuracy, and the master will lose calibration faster. On the other hand, when manufacturing to extremely tight tolerances, one might need a ratio of 4:1 or even 3:1 between gage and standard simply because the master cannot be manufactured and inspected using a 10:1 rule.

Take a taper master, for example. Say the angle tolerance is 0.001" over a 12-inch long taper. Usually a gage or master is not made 12 inches long. Rather, it may be 1 inch long. At this length the same tolerance now becomes 83μ". Using the 10:1 rule, the master would have to be 8.3μ". Unfortunately, this gage would be virtually impossible to manufacture or even measure.

But there are alternatives that can allow these tolerances to be measured and can reduce the cost of your masters. Masters can be certified to their class (the XXX, XX, X, etc.) or they can be certified to their size. What this means is that you may have the tolerance that requires a class XXX master ring. However, a suitable replacement might be a XX ring certified to size. What you will get with this ring is a Certificate that documents the ring's size at various locations and the calibration lab's measurement uncertainty.

Now you know that the ring met the XX class and you know the exact size of the ring. You can use this information to your benefit. When setting the gage to its reference (usually zero), set it to the actual master size. In effect, you are getting XXX performance from your XX ring. You've saved some money and probably sped up the delivery of your gage.

Wouldn't this be great if this worked for engagement diamonds also? That zirconium looks awfully good!

Section F

Holes

Economical Choice Of Bore Gage Depends Upon Your Application

Indicating bore gages come in two basic varieties: adjustable-capacity gages with interchangeable contacts or extensions; and fixed-size gages with plug-type bodies. While indicating plug gages can measure closer tolerances with higher repeatability than adjustable ones, these are only two of several factors to consider when selecting a bore gage. A wrong decision can mean unnecessary expense, low throughput, and even inaccurate data.

There is still a place in many shops for adjustable bore gages. Where tolerances are medium to broad, or production runs are low or involve many different bore sizes to be measured, adjustable gages can be a bargain. Their range is typically two to three times greater than that of plug gages (0.010" vs. 0.003" - 0.006"), so they are more practical to use with broader tolerances. Because an adjustable gage can measure a range of hole sizes, some shops can get away with just three units, with capacities of 0.500" - 1.00", 1.00" - 2.00", and 2.00" - 8.00". With indicating plug gages, on the other hand, a separate size plug is required for every different bore size to be measured.

The two types of gages are comparable in price (generally $400 - $600), but for a broad-tolerance operation, the smaller number of adjustable gages required will create a substantial savings. When you figure in the cost of masters, the savings can be multiplied.

For large-ID applications, adjustable bore gages are again the economical choice. Over about 4.5 inches, most plug-type gages are "specials" and, consequently, expensive (likewise for masters). Adjustable gages and masters are available from stock with capacities up to 24 inches.

The greatest benefit of fixed-size plug gages is the elimination of "rocking" to center the gage in the bore. The self-centering plug gage virtually eliminates operator influences and requires very little training. Rocking an adjustable gage is a refined skill that must be learned and performed conscientiously (Figures 1 and 2). A poorly trained operator, or one who is tired or hurried, is likely to produce incorrect measurements. Adjustable gages are also more subject to intentional operator influences, also known as the "close-enough syndrome."

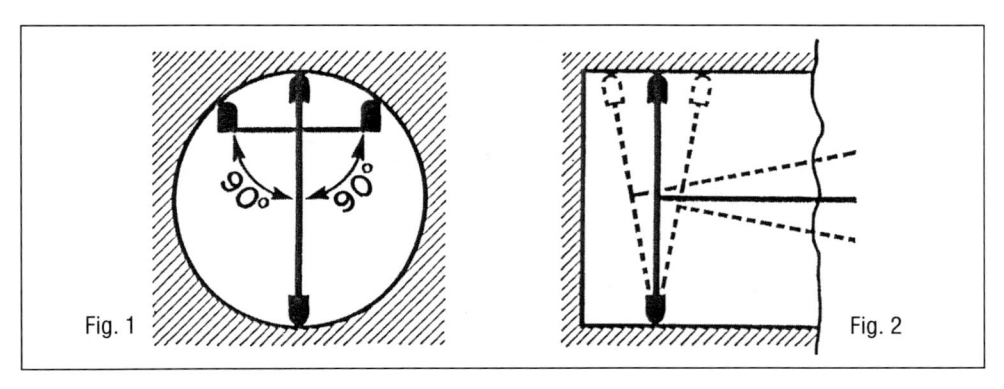

Fig. 1

Fig. 2

Left—Positive Centralization. Centralizing contacts locate the measuring contacts centrally on the hole diameter.

Right—Positive Perpendicular Location. Rocking gage slightly locates contacts squarely across hole as indicated by MINIMUM reading on indicator.

The elimination of rocking speeds measurement taking considerably. Mastering is likewise simplified and accelerated. In any production run where volumes are high and/or tolerances are tight, plug gages create time savings that quickly amortize their higher purchase price.

I am not endorsing indicating plug gages over adjustable ones, but the fact is, they offer more benefits overall. Plug gages have larger bearing surfaces, which make them less subject to wear. They are capable of better repeatability and discrimination. And they are the only logical choice for use with an electronic data collection system. It is nearly impossible to hold an adjustable gage steady on the true diameter of the bore, and at the same time, push a button to record the reading.

To summarize: Use adjustable bore gages where production runs are low and/or tolerances are medium to broad. Use self-centering plug-type gages where quantities are high and/or tolerances are medium to tight.

Measuring Deep Holes

Measuring IDs of "deep holes" involves a few special considerations. Deep is relative, but we'll define it here as being anywhere from roughly eight inches to 30 feet. Although the equipment and the methodology stay basically the same as for shallower holes, depth can influence accuracy, choice of gage type, and speed of operation.

The most basic deep bore gage is the mechanical, rocking-type gage, which comes in two basic flavors: self-centralizing and non-centralizing. The self-centralizing type is easier to use, because it requires rocking in only one plane. It offers high resolution (typically .0001"), but has very restricted range (typically .025" or less). The fixed contacts on the centering mechanism offset the gage head from the bore's centerline. In narrow deep holes this restricts the ability to rock the gage, which can interfere with some measurements.

The non-centralizing gage has no fixed contacts: it has two or three sensitive contacts that retract with a trigger mechanism. It requires a bit more skill to use, and has relatively limited resolution (typically .0005"), but a long measurement range (up to $1\text{-}^3/_8$"). Ironically, the head of the non-centralizing gage is automatically centered, which permits rocking to a greater extent.

Both types are available off the shelf for use at depths to about 12", but can be made to go deeper with the use of mechanical extensions. We've seen as many as six four-foot extensions screwed end to end to check for wear on the screw barrels of injection molding machines. However, because of the limitations of long mechanical transfer, it's difficult to obtain accuracy greater than about .0005" in this kind of situation.

Mechanical plug gages are self-aligning, which reduces the error potential of rocking-type gages and speeds gaging throughput. When used in conjunction with dial indicators, mechanical plug gages are subject to the same limitations of long-motion transfer as rocking-type gages. But it is possible to mount an LVDT right on the plug and feed them down the hole together on the end of a long rod. The signal from the LVDT runs over a wire to an amplifier on the "surface." In this way, mechanical measurements with resolution of .000050" are possible for the deepest holes.

Rocking-type gages are limited to IDs of roughly 1" or greater in deep-hole applications, while mechanical plug gages allow ID measurements down to about .200". To get smaller than that, you have to eliminate moving parts altogether and go to air gaging. With special hypodermic tubing, air gaging can measure down to .100".

Air is also practical for large IDs. It is often difficult to thoroughly clean a deep hole, and air gaging is very forgiving of dirt, oil, and other contaminants, both in terms of accuracy, and in terms of maintenance and longevity of the gage. This is especially important when gaging IDs of oil well pump barrels, which are some 30' long and about as dirty as anything you'd ever want to gage.

Air's non-contact aspect provides other benefits. It is common practice to measure IDs at two-foot intervals along the length of these pump barrels. Mechanical gage contacts would be subject to a great deal of wear under these conditions, but the jets on an air gage are unaffected. In another example, when measuring IDs of nuclear fuel rods, a non-contact gage is essential to avoid burnishing interior surfaces.

Because air gaging is, in a sense, the "standard" for deep holes, it's worth noting a few application tips:

- The deeper the hole, the longer air pressure takes to stabilize. In the case of an oil well pump barrel, this can mean 10-15 seconds.
- Air gages are calibrated for use within certain ranges of atmospheric pressure. If a gage is to be used at widely different heights above sea level (common for some oil-field users), it will be necessary to recalibrate it. This is easily done by checking a lookup table from the manufacturer.
- Special dial faces make some jobs easier. The oil industry, for example, commonly uses a face with +.008" to the left of Zero, and just -.002" to the right of it. This is because they are only checking for oversize caused by wear. Most gage makers will gladly design dials for special applications.
- Air plugs are available in a wide range of standard styles and specials. For gaging blind holes and counterbores, reverse venting is machined into the plug to allow air to escape. Blind- and super-blind plugs with very short lead-in sections are available for measuring to within .085" of the bottom of blind holes. Plugs may have pins inserted or dogs machined in to align the jets with specific workpiece features (e.g., for checking groove and land diameters on rifled gun barrels). Jets may also be arranged to permit measurement of TIR (total indicated reading) conditions and straightness as well as IDs.

In most instances, deep holes are just like shallow ones—only deeper. Measuring them is not a deep subject: it just requires a bit more care to select the right tool for the job.

Long Range Bore Gages

Long range bore gages offer users a great deal of versatility. These rocking-type gages without fixed centralizing contacts offer measurement ranges as "short" as $1/8$" (3 mm) and as long as 1 $3/8$" (35 mm). With replaceable contacts of varying lengths and shapes, their capacities are readily adjustable across a large span of nominal sizes: more than 2" (51 mm) of adjustability is not uncommon. Taking both their long measurement range and their adjustability into account, it is possible to inspect IDs from 0.670" to 7.462" (17 mm to 189.5 mm) with just four gages.

Long range bore gages feature trigger mechanisms that retract the contact points, permitting the gages to be inserted with ease in bores of varying sizes. This feature also provides access to bores that are larger than the entry hole. The contacts are sprung strongly to the "out" position, so releasing the trigger puts the contacts firmly against the sides of the bore.

These gages are normally equipped with long range dial indicators that typically resolve to 0.0005" or 0.001" (0.01 mm or 0.025 mm). In comparison, rocking bore gages with centralizers may have either dial or digital indicators, and typically have a shorter measurement range of about 0.025" (0.39 mm), but higher resolution of 0.0001" (0.002 mm). Thus, the long range gages are more appropriate for use in low to medium-tolerance inspection tasks.

Common applications include: gaging IDs of rough castings or forgings; measuring features whose nominal dimension is unknown; and inspecting used parts that have been subject to substantial amounts of wear. For example, IDs of used oil field pump barrels as long as 70' can be inspected with the use of extension rods. Of course, long range bore gages can also be used for conventional inspection of medium-tolerance ID machining operations.

Many long range bore gages are available in both two- and three-contact versions, the latter having the contacts oriented at 120° to each other. The two-contact gages are particularly flexible, and may be used in non-ID applications to measure inside dimensions between two parallel surfaces: slot widths, for example. The three-contact versions, which can only measure circular features, are useful for inspecting bores for three-point out-of-roundness.

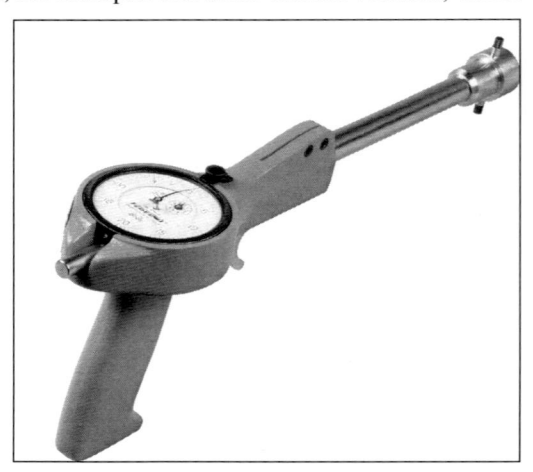

Long range bore gages must be centralized by "rocking" them in two directions. This is essential to ensure that the gage measures the bore's true diameter. First, the gage is rocked side to side, to center the contacts on the bore's axis. While rocking side to side, the user observes the indicator needle and watches for the maximum reading, indicating that the gage is not measuring a chord of the circle. Next, the gage is rocked up and down, to set the contacts perpendicular to the axis of the bore. In this case, the user watches for the minimum reading, indicating that the gage is not measuring a diagonal.

Non-centralizing bore gages require high levels of care and experience to obtain accurate results. A hurried approach will produce errors. The operator must have the patience to rock the gage several times in both directions while closely observing the indicator needle as it swings. Skill is also required to maintain the side-to-side centering while rocking the gage up and down. Long range bore gages with three contacts are somewhat easier to centralize than two-contact gages, but still must be rocked in both directions.

Other types of bore gages require less skill, and are quicker to use. But the long measurement range and adjustable capacity of non-centralized bore gages make them especially versatile, and consequently, potentially very economical where bores of many different nominal dimensions must be measured in small quantities.

Checking Bores For Ovality And Taper

Do you know the amount of ovality and taper of your bores? Are you sure you want to know? The decision can mean a big difference in the gage you select and the way you use it.

Checking ovality is primarily a job for air gaging, but there are exceptions: If you are interested in knowing whether a condition of ovality exists, but you don't need to actually measure it, then you might get by with an adjustable, rocking-type bore gage. Take one diameter measurement, then rotate the gage 90 degrees and take another. If the bore is oval, you will see a difference in the readings. But because you didn't necessarily hit the highest and the lowest points, you won't know how oval it is.

It is impractical to use a rocking-type gage to measure ovality, because it is virtually impossible to hold the gage in alignment while you rotate it through a full 180 degrees.

To measure ovality, you will want an air plug with two, and only two, jets, located 180 degrees apart. Take a measurement, then rotate the part (or the plug, depending on the setup) through a full 180 degrees, noting the maximum and minimum readings on the dial. (A two-contact, mechanical plug-type gage will also work for this application.)

If, on the other hand, you want to ignore ovality, select an air plug with four jets set 90 degrees apart. By the nature of air gaging, the four jets will average the readings between the minimum and maximum diameters. You won't even have to rotate the part on the plug. This is fine for some applications; for example, a press-fit bushing that conforms to the shaft when it is installed. In fact, it may be desirable for the purposes of process control to ignore the variations that would show up if one part were measured at its largest ID, the next part at its smallest, and the third one somewhere in-between.

The above considers only simple ovality—essentially, a two-lobed condition. If you wish to measure lobing of greater frequencies, air plugs with as many as 12 jets can be used. Checking bores for undesirable taper is similar to checking ovality. (We will not discuss measuring intentionally machined tapers.) Use a multiple-jet air plug to measure first near one end, and then near the other end of the bore, and simply note the difference, if any. If process analysis has indicated a need to check for barrel shape or bellmouth, take a measurement in the middle as well.

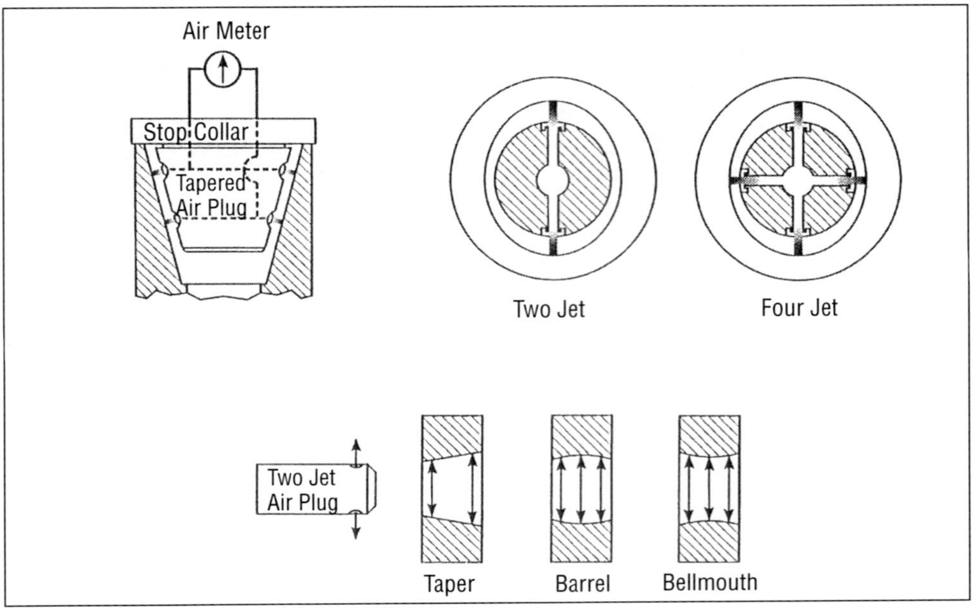
Air Meter
Stop Collar
Tapered
Air Plug
Two Jet
Four Jet
Two Jet
Air Plug
Taper
Barrel
Bellmouth

Of course, a four-jet plug will automatically ignore ovality, which may be desirable. If you wish to check ovality and taper in one operation, use a two-jet plug, measuring the bore at the bottom and the top, and rotating the workpiece or the plug at both ends. This method can get confusing, however, and you may prefer to do it in two separate operations.

Air gaging is a natural choice for measuring holes that are intended to be tapered (for example, Morse taper). This method of measuring taper is easy, fast and accurate, although it involves slightly more elaborate equipment. A plug with two separate air circuits is connected to the air gage so that each circuit acts on opposite sides of the precision diaphragm. Simply place the workpiece on the plug, and the gage will automatically indicate any variation in taper, based on the differential of air pressure between the two circuits.

Air gaging makes quick work of measuring ovality and taper. When to select air gaging over mechanical gages for other applications is the subject of "You Won't Err With Air, (page 168).

Measuring Blind Holes And Counterbores

Some time ago I wrote, that when measuring hole diameters, adjustable, "rocking"-type bore gages work well for broad tolerances and small production runs, while fixed, plug-type gaging is the practical choice for tight tolerances and high throughput. Plug-type gages automatically center themselves in the hole, eliminating the possibility of angular error in measurements. Let's now consider measuring blind holes and counterbores with plug-type gages. Both types of holes are gaged similarly, so for the sake of economy, I'll use the term "blind hole" throughout.

When machining blind holes, the front face of the tool does all the work, so it wears more quickly than through-hole tooling. Also, in any blind hole, a fillet occurs naturally in the angle between the "front" (i.e., bottom) and sides. As the tool wears, the fillet starts to inch up the sides, constricting the ID along the way.

It's easy enough to throw out bad parts when they occur, but a better reason to gage parts is to control the process, and avoid making rejects in the first place. In order to catch tool wear in time to control the process, special tools are needed for blind holes. Blind hole gages measure close to the front of a hole. How close? There are three types of blind hole gages.

Gage type	Height
(standard) blind	.156"
super-blind	.08"
super-super-blind	.030"

The height figure is the distance from the front of the plug to the centerline of the sensitive contact. (For comparison, the height-to-contact-centerline on a through-hole gage is $1/2$" to $3/4$".) The height specs, and the "blind/super/super-super" terminology are, believe it or not, industry standards.

Examine the part print to find the critical depth of the hole diameter, and choose the type of blind hole gage accordingly. Check especially if a precision mating part has to bottom out in the hole: an overly tall fillet will cause unacceptable interference. "Super" and "super-super-blind" plugs are usually used in connection with shallow holes, while standard blind hole gages may be for holes of any depth. All three types are available in diameters ranging from .217" to about 14".

Blind hole gages are relatively susceptible to wear and damage. The plugs lack the tapered, lead-in sections of through-hole gages, so the measuring surface itself is used to line the gage up to the hole. In the process, these critical dimensions are subject to a lot of bumps and wear. And the sensitive contact, being much closer to the front, can be readily struck against the edge of the hole; this is a common source of damage.

The entire front end of a through-hole plug could, in theory, be worn away, and the back end would still center the gage in the hole. With a blind hole gage, there is no front end: centralization occurs only behind the sensitive contact. Wear, therefore, begins on the only available centralizing surface from day one. Because blind hole gages are more subject to wear and also less tolerant of it, one cannot expect them to have the longevity of through-hole gages. So don't use blind hole gages where through-hole gaging would work.

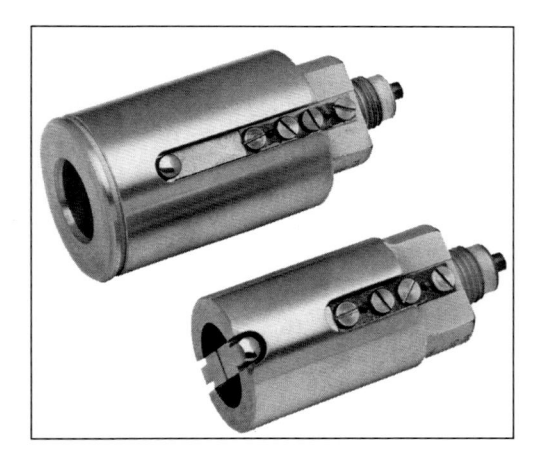

Although a blind master is recommended, many people use standard master rings to check blind hole gages. You can get away with this if you're careful.

Make sure you master at the same depth you measure; otherwise, wear at the front of the gage body may be masked by better centralization further back. This could give you great repeatability on the master, and none in actual practice.

Also note that master rings are only certified at certain depths: there's no guarantee they're accurate near the top or bottom. Be especially wary of this with super-blind gages. Drop a few quarters in the ring to bring your gage up to the right height. The key to good mastering is to duplicate measuring conditions as closely as possible.

Through-holes tend to be self-cleaning, but blind holes need to be cleaned of chips and other debris before measuring. One can use coolant if the hole faces downward. Otherwise, compressed air is the best bet.

■ ■ ■

A Shallow Bore (This Is NOT An Autobiography)

Imagine a large rotor with an ID of 12". One normally uses an adjustable bore gage, or perhaps an inside rod micrometer, to check the diameter. But in this case, a hub in the center of the part presents an obstacle. For this type of measurement we need a special type of gage, called a shallow bore gage (sometimes called a shallow diameter gage or shallow ID/OD gage). These gages may be used where the diameter is reasonably close to the face surface of the part (usually 3" or less), and where the face surface is flat and square to the feature axis. While some of these gages can measure diameters as small as 2" or so, larger dimensions are more common, and gage capacity can range up to several feet.

A rail usually serves as the frame of a shallow bore gage. An anvil post, with a reference contact, extends perpendicularly from the bottom of the frame. The sensitive contact extends below the rail at the opposite end, for the final leg of the "C." The sensitive contact is connected via a low-friction linkage to an indicator on top of the rail. Depending upon the specific gage, either or both contacts may be adjustable along the rail, and both are adjustable for depth. (For gages offering more than 3" of depth capacity, special bracing or heavier contacts are required to provide the necessary rigidity.) The contacts are usually back-tapered, to ensure true point-to-point measurements. Rest feet are provided to establish a reference plane parallel to the measuring plane of the part. These, too, are adjustable linearly to accommodate different diameters.

Most gages offer about 6" of capacity adjustment. On some gages, rails of different lengths may be interchanged, so that the range of adjustment may be virtually unlimited. Frames can be configured to avoid interference with protruding features in the center of the part. With so many adjustments possible, shallow bore gages are adaptable to a wide range of applications, including grooves, tapers, cylinders, and features recessed behind blind shoulders.

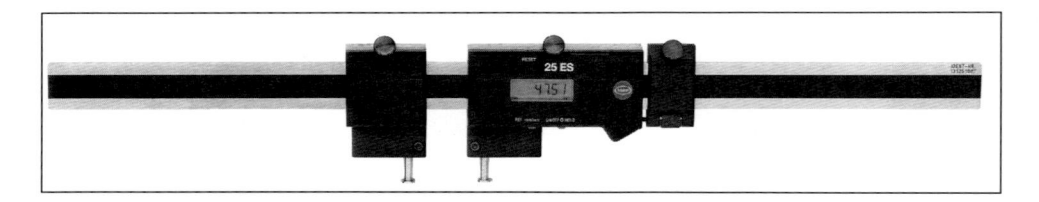

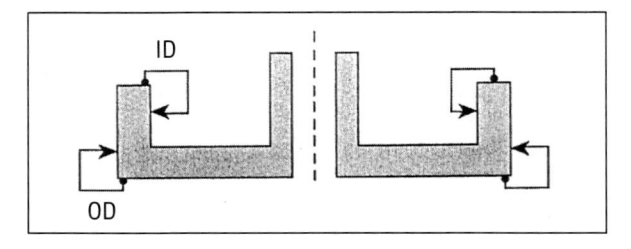

Because the diameters being measured are usually large, dial indicators with resolution of 0.0001" or 0.0005" are typical. For applications requiring higher resolution or data output, higher resolution indicators, digital indicators, or even electronic transducers can be substituted. The electronic devices, with their ability to automatically capture the Min or Max value during the sweep, may also be preferred as a means of reducing operator errors. Some gages provide the option of mounting the indicator either horizontally or vertically, to improve visibility in some applications.

In use, the gage is swept across the diameter to find the maximum or minimum reading, depending on whether the feature is an ID or OD. Some gages include a centralizing device, which can be helpful but does not eliminate the need to "rock" the gage.

Mastering is usually done with adjustable setting masters, which must first be set to the nominal dimension with a gage block stack or end rods. Gaging depth is usually set by eye, placing a steel scale beside the contacts as they are screwed in or out. Where depth is critical (as with tapered features), the rest feet are placed on gage blocks of the appropriate height, and the contacts are adjusted up or down until they just touch the reference surface. If the bottoms of the contacts are radiused, make sure the gage block height is raised accordingly.

■ ■ ■

Gaging Countersunk And Chamfered Holes

While countersunk and chamfered holes are similar in appearance, functionally they are quite different. Consequently, different gages exist to serve these different functional requirements.

Hole chamfers are usually specified simply to make it easier to insert a screw, pin, bushing, or other assembly component. After the part is assembled, the chamfer typically serves no function. The component doesn't bear on the chamfer, so diameter and angle tolerances are usually not critical to the part's performance.

A countersink, on the other hand, is a functional surface upon which a fastener head bears. Because fastener performance is so important, countersink tolerances are critical. The countersinks on an aircraft's skin are an excellent example. If the countersink is too deep, there may be inadequate skin material for the rivet to hold against the underlying frame. If it is too shallow, the rivet head will protrude, increasing air resistance. This latter condition may sound trivial, until you consider the cumulative effect of literally hundreds of thousands of protruding rivet heads on an airplane's skin.

Countersinks tend to be small—usually 0.780" or less—and angles are closely controlled: usually 30°, 82°, 90°, 100°, or 130°. Chamfers may be specified at any angle up to 130°, and on holes or inside diameters of any size.

Both countersink gages and chamfer gages are usually hand-held instruments, although both can be mounted on bench stands, which is convenient if the parts being measured are small. They both perform the measurement by means of a plunger mechanism, but they do not measure the depth to which the angled surface extends into the hole. Rather, they measure the major diameter of the feature—that is, the largest diameter of the hole, where it intersects the top surface of the part. To convert the vertical motion of the plunger into a diameter measurement, the gages require an indicator with a special-ratio movement or readout.

The main difference between countersink and chamfer gages is the configuration of the plunger. Chamfer gages have an angled plunger consisting of three fluted sections. The angle of the plunger must be greater than the angle of the chamfer, to ensure that the plunger contacts the major diameter only. Typically, the range is split in two—from 0° to 90° and from 91° to 130°—so that two gages have traditionally been required to measure the entire range of chamfers. Recently, however, gages with replaceable plungers have been introduced (made possible by the introduction of electronic indicators with adjustable-ratio displays), allowing a single gage to be used with chamfers of any angle. Replaceable plungers also provide flexibility to measure across a larger range of diameters, and to switch between ID and OD chamfers.

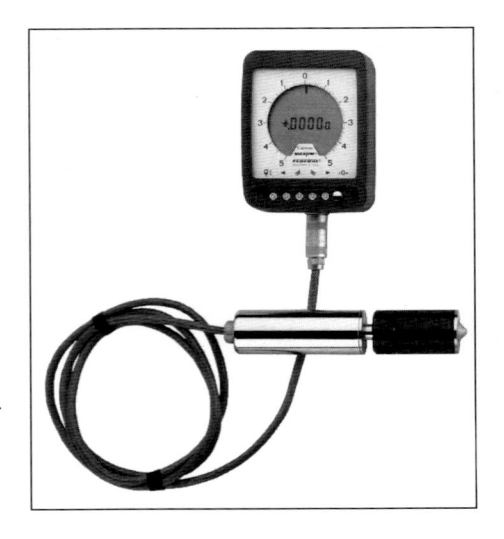

Because countersinks are more critical, countersink gages have conical plungers that fit closely against the entire surface of the countersink feature. While some slight difference in the angles of the plunger and the countersink is acceptable, there must be a fairly close match between them. There are no replaceable-plunger countersink gages available at this time, so separate gages are required for each angle requiring inspection.

Both types of gages may be mastered against a part master that duplicates the specified chamfer or countersink. Chamfer gages can also be mastered against any certified flat surface. In most cases, this means the gage will perform like an absolute or direct-reading gage, in which case the indicator displays the feature's actual diameter. Alternately, if the chamfer gage has a digital indicator that allows pre-sets to be entered, then it can still be used for comparative (i.e., plus or minus from nominal) measurements, even if it is mastered on a flat.

■ ■ ■

Setting Adjustable Bore Gages

Adjustable bore gages are comparative type instruments. This means that the readout, whether it be a dial or digital indicator, or other type of electronic readout, will show the amount and direction of variation in the test bore from nominal size. That being the case, the gage needs to be set to the required nominal size to which the actual bore is to be compared.

The gaging components of an adjustable bore gage are mainly in the head of the gage. These include a sensitive contact through which the measurement is transferred mechanically to the readout device; a pair of centralizing contacts that align the gage radially to the bore; and a reference contact that is used to set nominal size. Setting the bore gage consists of adjusting this reference contact to produce a zero reading on the indicator when the gage is measuring the nominal size.

There are a number of ways that the adjustable bore gage can be set to the nominal size. Some are good for a quick setup, some are more expensive but very precise, and some provide a good balance of reliability and versatility.

Using a gage block assembly in a clamp with jaws at both ends provides a flexible and highly accurate reference master.

Using an outside micrometer may be the quickest and most available method. But there are issues with the accuracy of this setup. Micrometers have inherent errors that can be passed along to the gage. Another big issue is accurately locating and aligning the spherical measuring points of the gage on the contacts of the micrometer. While readily available and inexpensive, this setup method probably offers accuracy of no better than 0.002". This accuracy must be compared to the part tolerance to see what portion is consumed by inherent error.

Probably the best method for setting the adjustable bore gage is with a master ring. This duplicates the actual measurement, and master rings can be made very close to the part size. Most bore gages have a flat surface on the head that is parallel to the reference and sensitive contact. Using this method, the ring is laid on a granite surface plate, the bore gage is set in the ring which will support it, and the sensitive contact is adjusted until the indicator reads zero. It is even possible to use a ring gage that is not exactly the nominal size, as the offset can be incorporated into the set of the dial indicator. A ring gage is the preferred method for setting repetitive sizes or when an adjustable bore gage is going to be dedicated to a particular size. The downside of this method is the potential cost. Rings can be expensive and if one is required for each of many sizes on the shop floor, the total cost may be prohibitive.

Using a gage block assembly in a clamp with jaws at both ends will also provide a highly accurate reference master. When multiple sizes are required and flexibility is key, this can be the preferred method. Gage blocks are the most basic reference standard. They are readily available and provide high accuracy. The only drawback is in the time required to assemble the gage block stack to the nominal size. Also, since only a single stack is used, there is no verification that wringing errors have not been made in assembling the stack. Though small these could affect the performance of the gage.

As with setting any gage to its nominal size, care needs to be taken. Setup of the masters or gage blocks is key. This means they must be clean and free of dirt. Also, they need to be temperature stable, as does the gage.

Once the reference contact is set to the desired size, the locking screw should be made snug, and the gage tested again for zero and repeatability. If nothing has changed, keep snugging down the locking screw until it is locked into position. Then check repeatability again until you are confident that the gage is performing accurately.

Both the master ring and the gage block method provide the best setting performance, and can provide overall accuracies of 0.0001" to the nominal size.

Section G
SPC

Gaging For SPC: Keeping It Simple

After all that has been written about Statistical Process Control (SPC) in the past few years, I am continually surprised at how often shop owners tell me they would like to do SPC, but they're "just too small," or they "can't afford all that fancy equipment." There seems to be a mistaken yet growing perception out there: 1) that SPC is a lot more complicated than it is, and 2) that you need a computerized system on line just to get into it.

This is unfortunate because small shops can often benefit the most from SPC and there really is no reason they should not. The simple fact is, you don't need a computer and you don't need to be a statistical genius to do SPC. I know this latter for a fact, because I understand it. All you need, really, is an indicating gage, a pad and a pencil. So if you're one of those who is still hesitant about charting, try this easy-to-follow recipe for small shop SPC:

Before you start, you do have to understand a little bit about the process itself. But this is not difficult. Bear in mind that the basic principles of SPC were developed back in the '30s—long before computers were invented—and have not really changed since. So don't be intimidated by all the bells and whistles. An X/R chart then was the same as an X/R chart now. You don't need to understand all the theory behind the process, just the basics of frequency distribution and charting will do for a start, and there are many guides to help do this. (In fact, Federal Products published one back in 1945 which is currently in its 14th printing and is still in use!) What you're basically looking at is taking a few simple measurements, averaging them, then recording the results.

Next, get the process in place. This involves three steps. First, look at the part you're going to measure. What are its important dimensions and what part of the process controls them? Here is what you want to measure and where you want to start the process. But keep it simple. Start with a single measurement until you get the feel for it.

Second, look at the gaging equipment you are going to use. Make sure you follow all the gaging basics we've talked about in this column. Make sure it is the right gage for that type of measurement, and that you follow the ten-to-one rule for resolution (to measure 0.001"

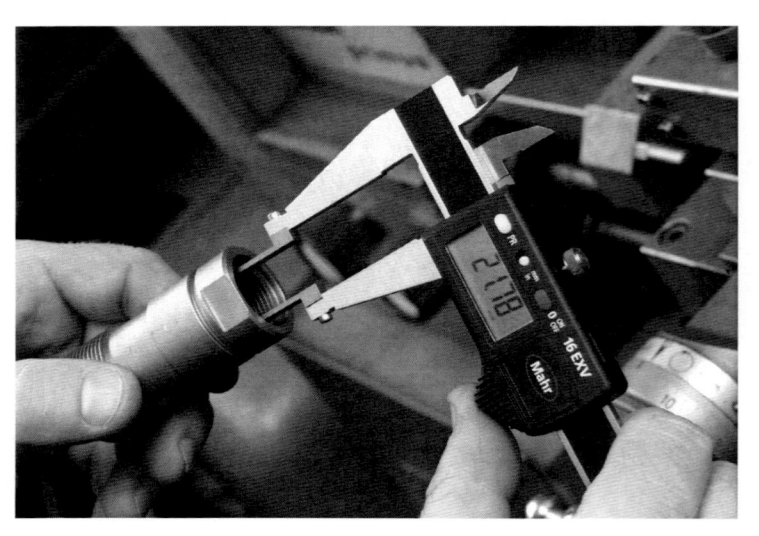

Most trainers agree that starting simple is smart when it comes to SPC. A digital indicator that records data electronically can ease the tedium of manual charting and eliminate errors, while still keeping the operator involved, by giving him an immediate reading so he can monitor his process.

tolerance, you need a gage with at least 0.0001" resolution). You should also run a series of Gage Repeat and Reliability (GR&R) studies to make sure your measurements are as accurate as possible. Remember, whatever "analysis" you do can only be as accurate as the measurements you start with.

The final step in setting up is to benchmark your machining process. You need to find out, simply, if your machine is capable of holding the tolerances or control limits you require. Your SPC guide can help you do this.

Now you're ready to gather data. There are many sophisticated systems available to facilitate this process. However, most trainers agree you are much better off starting manually. With automated data collection systems, the operator often feels left out of the process: it happens without him. He doesn't need to think about it, and therefore makes no effort to understand it. This is bad because, ultimately, it's not SPC that controls your quality, it's the operator. If he doesn't understand the process, he is unable to use it.

Charting manually, on the other hand, puts him immediately in contact with the process. He is able to see that there is a relationship between what he's doing and what the charts show. He sees that his process changes and that the charts provide a prediction of how it changes. He is then able to control it. He sees his wheel is wearing, for example, and knows when, and when not, to compensate. Charting manually—or at least visually—empowers the operator, and that, ultimately, is what SPC is all about.

The other advantage of manual charting is that you can use just about any type of indicating gage, so long as it is accurate. Since you probably already have some in your shop, your SPC investment cost is reduced to the Manual, the paper and the pencil. Who knows, you may even already have the pencil.

On the downside, manual charting is tedious and time consuming, and subject to error. Whenever an operator measures and records manually, he has two chances for error: he can observe wrong and he can record wrong. This is where you may want to consider a digital indicator, which can give you several advantages. First, you can eliminate error by gathering and recording data electronically—as well as interfacing with whatever computerized system you may want to put online. Second, it makes it quicker and easier to collect the readings: all the operator has to do is measure and press a foot switch. Battery operated gages provide a back-up in case your computer system does go down and, most importantly, these gages still provide the operator with an on-site reading so he can monitor his process.

That's it: simplified SPC for the small shop. Start small, take it one step at a time, and keep your operators involved. And remember, SPC is not a thing you buy, it's a thing you do.

Bedrock S.Q.C.

Statistical Quality Control has been in use since the 1930s, when it was performed with paper, pencil, and maybe a slide rule. As such, it was originally somewhat time consuming, and it required a fairly high level of care and understanding on the part of its practitioners. SQC got a big boost in the 1970s and '80s, when electronic gages, data loggers, and PCs began to proliferate. Suddenly, untrained line inspectors could easily perform the neces-

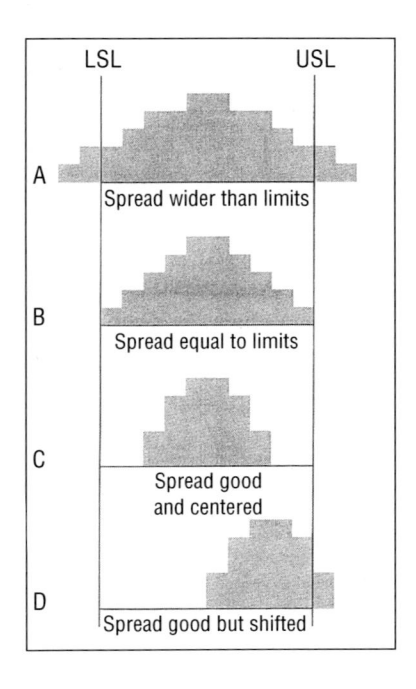

[Figure: Histograms labeled A, B, C, D between LSL and USL markers:
- A: Spread wider than limits
- B: Spread equal to limits
- C: Spread good and centered
- D: Spread good but shifted]

sary calculations, without really understanding the process.

Many instructors still believe, however, that inspectors and machine operators should learn to do SQC manually before they plug in their data loggers, on the general principle that people who understand what they're doing tend to do a better job. Thus, we'll be looking at some of the basics of SQC in the next column or two. The rules can apply to any dimensional gaging procedure in a high-volume production application.

All operations that produce features to dimensional tolerances involve variation. Variation cannot be eliminated, but it can be controlled so that it remains within acceptable limits. SQC uses the laws of probability to reliably monitor and control a process. By inspecting variation in a small sample of production, it is possible to draw inferences for the entire lot.

Let's use the following simple example: the feature to be inspected is an OD on a rod, and the specification is 0.375" ±0.005". For our sample, we select 35 parts at random from a certain segment of a production run. (Let's not worry, for the moment, about how we arrive at that figure. Suffice it to say that samples need rarely be greater than 50 parts.)

The first "statistical" procedure is to find the smallest and the largest measurements, then subtract one from the other. This is the "range," which we might express as R = 0.008". Next, we find the average of all the measurements (the sum of the measurements divided by the number of measurements in the sample), which we express as the "X-bar" value, as in X = 0.377".

These very simple procedures provide important information. First, they tell us whether all the parts in the sample are in tolerance, and how much of the tolerance range (0.010") the variation (0.008") consumes. And secondly, they show the relationship between the average value and the specification. The average of a sample may lie exactly on the part specification, but the range of the sample may be broader than the tolerance range, so that many parts lie beyond both the upper and lower tolerance limits. On the other hand, the range of the sample may be smaller than the range of the tolerance limits, but the average may be skewed so far from the specification that the entire sample falls outside one of the tolerance limits.

The next step is to chart the data on a histogram. A range of measurement values is divided into equally spaced categories, and each measurement is placed in the appropriate category. If distribution is "normal", the resulting Frequency Distribution Curve will have the familiar bell shape. The "mode"—the category containing the largest number of data points—will be the same as the average in a normal distribution. Distribution curves that are not bell-shaped may indicate a problem in the manufacturing process. For example, if the curve shows a dip where the mode normally appears, it might indicate looseness in the setup or the machine tool.

The histograms show four examples of normal distribution. In A, the range is too wide, indicating that a large percentage of production falls outside the tolerance limits. In B, the range is equal to the tolerance limits: all the items in the sample pass inspection, but there is a statistical probability that some parts in the production run will exceed the limits. In both cases, one would want to reduce the range of variation.

The range of curve C is significantly less than the tolerance range, and falls entirely within the specifications: there is a probability of very few bad parts in the run. In D, the range is acceptable, but because the average is displaced to one side, a significant number of sample parts fall outside of tolerances. Some means must be found to shift the average while maintaining the range.

■ ■ ■

More On Gaging Statistics

A histogram, showing dimensional variation in a random sample of parts, can be used to determine whether a process is under adequate control. Both the range (R) and the arithmetic mean (X) of the sample must be monitored to make sure that parts consistently remain in tolerance. Let's continue on this topic. Deviation, as a statistical term, tells how much a given piece of data diverges from the mean. (For example, if the mean value of a sample is .010", and the part in question measures 0.012", deviation is +.002".) Standard deviation, when designated as S, is a measure of how much the values of all the individual items in the sample diverge from the mean value of the sample. Standard deviation can also be designated as σ (the small Greek letter sigma), which is an estimate, based on a sample, of how much the values of the individual items in the total population from which the sample was drawn will diverge from the mean value of the population (see figure).

We won't go into how to calculate S or σ here (the methods appear in any basic textbook of statistics, or you can just punch up the values on a scientific calculator). Suffice it to say that once you have found the value of σ, further standard deviations are, by definition, simply multiples of σ. (For example, two standard deviations = $2 \times \sigma$.) Some of the data in a sample will fall beyond one standard deviation from the mean. The second standard deviation tells us how much additional deviation exists among those parts that do not fall within the first standard deviation; the third standard deviation tells how much deviation exists among parts that don't fall within the first two deviations, and so on.

Standard deviations can be displayed on a distribution curve as pairs of positive and negative bands on either side of the mean value, as shown in the figure.

In any random sample showing a normal, bell-curve distribution, one standard deviation (that is, one positive and one negative band) will encompass roughly 68 percent of the values in the population; two standard deviations will encompass roughly 95.5 percent; and three standard deviations will encompass roughly 99.75 percent of the workpieces. It always works out this way, because of the laws of random distribution and statistics: wider bell curves naturally exhibit larger standard deviations, and narrower bell curves exhibit smaller ones, and the two are always in proportion to one another.

By comparing the width of the standard deviation bands with the specified tolerance limits, it is possible to calculate, from the sample, how many bad parts will be produced for the production lot from which the sample was drawn. For example, if the tolerance limits are equal to plus or minus three standard deviations, and the mean of the sample is perfectly centered on the specification, one could expect 25 bad parts out of every 10,000 produced (100 percent - 99.75 percent). It is simply a matter of calculating the mean and three standard deviations, and comparing the results against the tolerance specification. If the "± 3 sigma" spread falls entirely within the tolerance limits, as in the figure, then the process appears to be under control. If part of the spread falls above or below the tolerance limits, then the process must be adjusted to make the bands narrower (that is, reduce variation), change the location of the mean, or both.

In any precision manufacturing operation, dimensional variation is very tightly controlled, and the sigma limits are monitored to keep them within the tolerance limits. The span can be compared to the tolerances in various ways. This is known as process capability.

In summary: it is possible, based on the laws of statistical probability, to effectively monitor a process and maintain high levels of quality control using a sampling method. In most applications, this tends to be far more cost effective than 100 percent inspection. Rather than drawing histograms, one can calculate the "control limits"—that is, the upper and lower boundaries of the standard deviation bands, and the acceptable range of variation of the mean—mathematically.

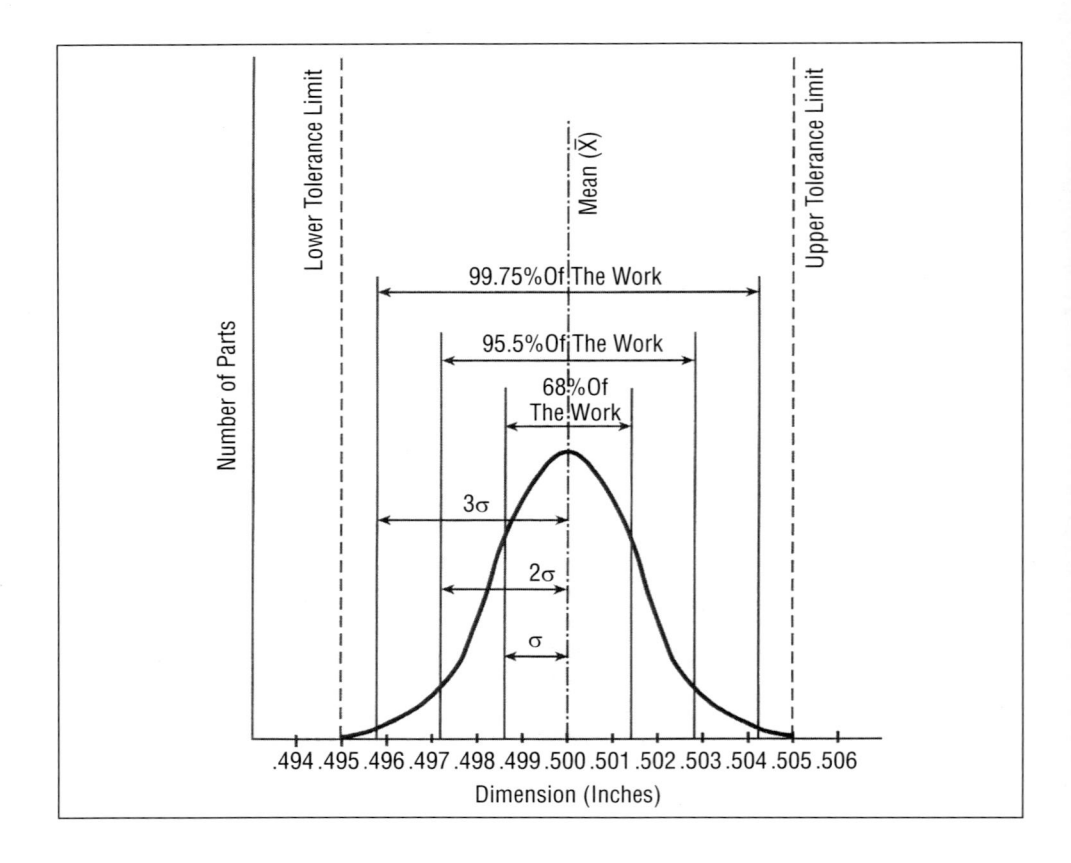

Assessing Gage Stability

When multiple part measurements show unacceptable variation, it is essential to understand whether it is the manufacturing process or the measuring process that is at fault. If inaccurate measurements are relied upon, one may make adjustments to a manufacturing process that is in reality accurate and under control.

Gage stability implies different things in different contexts. If taken literally, it may refer to whether there is something loose on the gage, or some other gage problem occurs randomly, to cause two identical trials to produce different results. When plotted on a histogram, this type of instability shows up as a distortion of the expected bell-curve shape—perhaps in the form of dual modes (i.e., high points) with a dip in-between, or in a mode that is skewed toward one end of the tolerance range or the other.

Other types of stability are best assessed and visualized through the use of control charts, in which measurements of several small lots, each represented by a histogram, are compared over a period of time. A single, certified master or other qualified part should be used for the entire series of repeated trials. This serves to eliminate part-to-part error as a variable, so remaining variation will likely be low. Nevertheless, testing may show that variation between lots becomes significant enough over an extended period so as to constitute a measurement problem.

"Statistical stability" refers to the consistency of the measuring system's performance from lot to lot. If a gage is statistically stable, each lot or histogram will be nearly identical in shape and range (R), and the average value (X-bar) of subsequent lots will be close to one another. If it is statistically unstable, the histograms will vary in terms of their X-bar or R values, or both.

Figure 1 shows statistical instability of both sorts. The histograms vary considerably in their ranges and shapes, while the mean dimensions are clearly irregular from lot to lot.

"Long-term" and "short-term" stability are confusing, because both require a long-term effort to assess. Both describe trends in the X-bar or R values across multiple lots. If we were to draw a line connecting the X-bar value of each histogram in Figure 1, we would observe no clear trend. Thus, we cannot make a meaningful assessment of either long- or short-term stability.

In Figure 2, the R values are nearly identical, and the X-bar values are grouped much more closely, so the gage is statistically stable. A line connecting the X-bar values would show a clear trend along the nominal value, so we have good long-term stability. This chart shows realistically how the performance of a good, stable gage looks.

Figures 3 and 4 both show good statistical stability. But in Figure 3, there is a short-term stability problem, with a bias in the plus direction. After a few sets of trials, however, the measurements become stable and remain so over the long term. The problem may be due to a warm-up condition in the gage itself. Or it might be the result of contamination that has settled on the gage over night. The X-bar value shifts as the contaminant is dispersed, due to mechanical interaction between the gage and the samples being measured. After a few dozen trials, all the contaminant is dispersed and the measurements become stable.

Figure 4 illustrates the opposite situation. The gage is stable on startup and over the short term, but over the longer term it becomes unstable as the X-bar values begin to drift

Section G: SPC

1.	2.
3.	4.

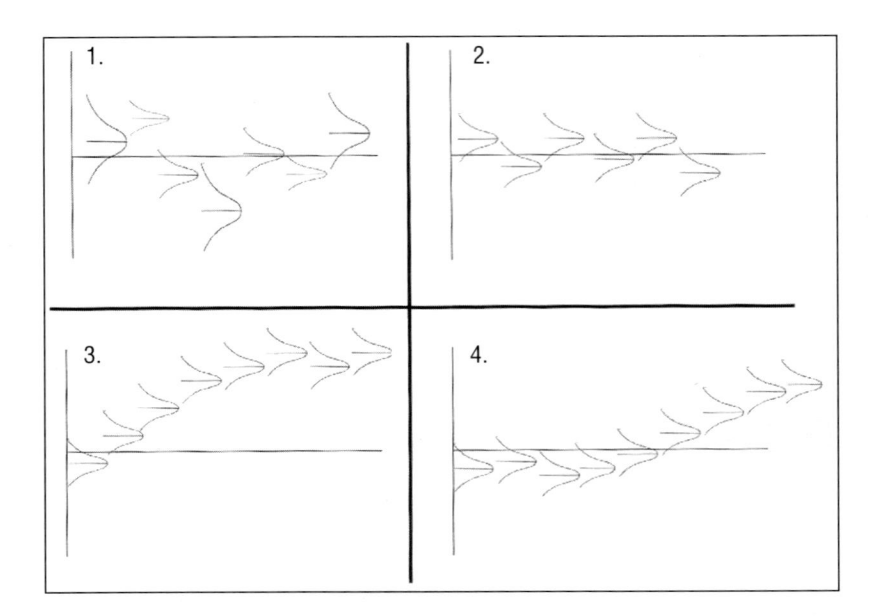

upward. This might be the result of external thermal influences. If performance were to be charted over a longer period, the X-bar values might drift back toward nominal again; this might indicate the gage is responding as the plant warms up in the morning, then cools down toward evening.

It is important to check gages for stability, in order to avoid making inappropriate adjustments to the manufacturing process. In some cases, solving the problem may be as simple as establishing a more frequent schedule for remastering the gage. In others, it could require a reevaluation of the entire gaging process.

Section H

Machine Calibration

Calibrating Machines For Quality

Machine calibration is often seen as too complicated and too time consuming to be worth the bother. And many shop owners reason, if the factory can't set a machine right, how can they? So they continue to purchase machines without a thorough check of their full range of motion, they set them up and run them—sometimes for years—without re-calibration, and then continually complain about their "inability to hold tolerance." If this description makes your collar pinch, here are some things to think about:

First and foremost, set at the factory does not mean set in your shop. Too many things can happen during shipment and installation to even hope a new machine will be in spec without calibration. Second, once set does not mean always set. Strange things can and do happen to machine tools: an errant shaft of sunlight heating up a lead screw, or a minor lubrication problem in the ways, can throw a very expensive machine all out of position. You need to recheck that machine and its environment on a regular basis to ensure its ability to produce good parts. The good news, though, is that with the advanced equipment and software available today, calibration is not the bear it used to be. And most important, an ongoing calibration program in your shop can pay hefty dividends.

The most obvious of these is better quality parts. This means fewer rejects, reduced scrap, and less rework. But, there are other, more subtle, benefits as well. One of these is in part and program editing. One of the "miracles" of the CNC revolution is supposedly the ability to program part routines and run them on different machines. This sounds good in theory, but in practice, it usually requires untold hours of programmer and operator time editing routines to accommodate machine peculiarities. Calibration minimizes this programming and editing time; so one tape really can serve several machines. This can make a major difference in the ability of a shop to respond in a JIT environment.

Another often-overlooked benefit is the ability to document quality. This is valuable not only for vendor certification programs, but also as a marketing tool to help sell your capability. But the most important benefit of an ongoing calibration program is the increased understanding you gain about your machine's performance and your overall production environment. Calibration not only tells the good from the bad, it tells you <u>how</u> good your machines are, and how you can make them better, <u>where</u> they are best, and <u>when</u> you can expect them to give you trouble.

This knowledge can pay off in a number of ways. Scheduling, for example. Knowing in a very precise way what your machines are capable of will not only help you optimize production, it can also help you do things you didn't know you were capable of. Sometimes machines have "sweet spots," ranges in which they perform way beyond their stated accuracy specification. If you know where they are, you may be able to take very profitable advantage of them. Maintenance and troubleshooting are another. Machines usually don't break overnight. There are warning signs. Monitoring performance on a regular basis can put you in the driver's seat. You will know when readjustments are necessary. You will be better able to schedule regular maintenance. And, you stand a better chance of being forewarned of major problems and avoiding the inevitable, middle-of-a-rush job breakdown.

Finally, regular calibration can help you determine when a favorite old machine has, shall we say, passed its peak. And when the time comes to replace it, the understanding you will have gained from ongoing calibration will make you a much wiser and better buyer. And help you prove it.

'Round And 'Round She Goes...

The calibration of machine tools is no longer an uncommon practice. By measuring the positioning and geometric contouring accuracy of a machine tool, it is possible to make effective adjustments, schedule preventive maintenance, and tighten machining tolerances. As a result, the use of calibration tools such as ball-bars, laser interferometers, and electronic levels is growing in popularity in machine shops.

In order to tweak out the last source of machine error, we have to look at the spindle itself. It is, however, difficult to take measurements on a spindle that's running at several hundred or several thousand rpm, unless you have the right tools. That's where spindle analysis equipment comes into play.

Spindle errors may be present, individually or in combination, in three different directions: radial, axial, and tilt. Radial error motion may appear either as synchronous or asynchronous errors. Synchronous error is a deviation that occurs fairly repeatably on every revolution of the spindle shaft, and it shows up on the workpiece as errors of form or geometry. Asynchronous error is non-repeatable, and it shows up as problems of surface finish. Axial errors tend to generate surface finish or flatness problems, while thermally induced tilt errors contribute to location problems in parts. The ability to measure errors can give the user a window on how the spindle contributes to geometry, position, and texture problems. By addressing spindle error, it is possible to reduce or eliminate these problems, which makes it easier to meet dimensional tolerance specifications.

It is critical to measure deviation under dynamic conditions, with the bearings warm and all normal sources of vibration present. Spindle error analyzers rely on capacitance gages to measure deviation while spindle is actually running, at speeds up to and over 100,000 rpm. "Cap gages" use the principle of electrical capacitance to measure the volume of air between the probe and the target—a precision steel test ball or mandrel fixed in the tool holder.

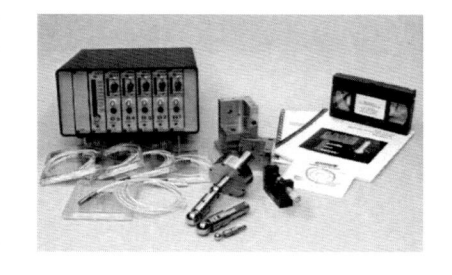

One or more probes may be held in a "nest" fixture fastened to the machine's table: with five probes in place, it is possible to measure radial, axial, and tilt errors with the same setup. (Assuming a vertical-spindle machine tool, the probe layout would be: two vertical pairs oriented at 90 degrees to each other, and one directly beneath the mandrel.) The procedure is described in the ANSI/ASME B5.54 standard, "Methods for performance evaluation of computer numerically controlled machining centers." In spite of its name, the methods are equally applicable to manually controlled machine tools. B5.54 also describes the use of a "wobble plate"—in which the steel test ball is intentionally offset from the spindle axis—as well as methods to test for short- and long-term thermal stability.

As part of the "toolbox" of calibration tests, spindle error analyzers play an important role in characterizing a machine tool's total potential accuracy and monitoring changes in accuracy over time: this is a valuable form of information for the efficient scheduling of jobs and machines. Spindle error analyzers are useful for checking out a spindle after a

Section H: Machine Calibration

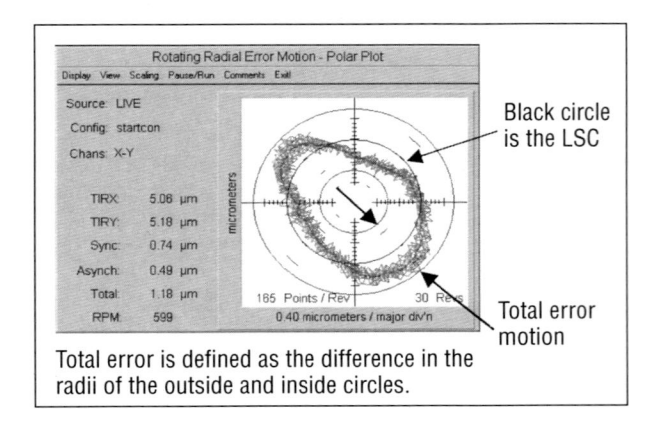

```
Rotating Radial Error Motion - Polar Plot
Display  View  Scaling  Pause/Run  Comments  Exit

Source:  LIVE
Config:  startcon
Chans:  X-Y

TIRX     5.06 µm
TIRY     5.18 µm
Sync     0.74 µm
Asynch   0.49 µm
Total    1.18 µm         165 Points / Rev    30 Revs
RPM      599              0.40 micrometers / major div'n
```

Black circle is the LSC

Total error motion

Total error is defined as the difference in the radii of the outside and inside circles.

crash—often a much cheaper approach than to simply resume cutting metal and hoping that nothing's wrong. They can serve as troubleshooting tools, to help identify problems of form and texture, and even to track down specific components in the spindle that may be causing the problems. And spindle error analyzers can be used for acceptance testing of new and rebuilt spindles, helping buyers make informed purchasing decisions by providing a means to objectively compare the accuracy performance of different makes and models.

There are three ways to observe spindle errors. You can see them in part rejects. You can hear them as noisy, worn spindle bearings. Or you can observe them before they become a real problem with a spindle error analyzer.

■ ■ ■

Is Your Machine Square?

We've looked at many important aspects of the performance of machine tools, including straightness of travel, pitch, yaw, and roll, and spindle accuracy (i.e., radial and axial deviation). Another critical aspect of machine geometry is that of squareness between axes. Any out of squareness will be directly reflected in the part being machined. And like pitch, yaw, and roll, the farther the machine moves from the intersection of its axes, the larger the influence of the squareness error.

Squareness can be measured with several different sets of tools, including: mechanical squares; telescoping ballbars; electronic levels; straightedges with indexing tables; and lasers with optical squares and straightness interferometers. The method one chooses depends upon the size of the machine, the accuracy required, the skills of the technician, and of course, cost (or what's available in the tool crib). The general guide to the measurement of machine tool squareness is to be found in the ANSI/ASME B5.54-1992 standard. By any method, measuring squareness comes down to comparing two nominally straight axes for a 90° relationship. We'll look at a few of them here.

Precision mechanical squares are a very economical way of checking a small machine—up to roughly 24" per axis. Beyond that length, the accuracy of artifacts tends to be inadequate for the purposes of machine evaluation, and/or their price becomes so steep that other methods become more economical. Mechanical squares come in a variety of configurations and materials, and are supplied with certificates of calibration, stating their maximum errors of straightness and squareness.

To measure the squareness of a vertical axis to a horizontal axis, the square is placed upright on the machine, and an indicating device (i.e., a test indicator or electronic gage head) is attached by means of a bracket or tool holder to the moving component of the vertical axis. The indicator is traversed across the vertical reference surface of the square, and measurements are recorded at convenient intervals. The data is plotted, and a best-fit line is calculated, the slope of which represents the squareness error (after allowing for the calibrated accuracy of the square itself). Like straightedges, squares are self-checking devices. To eliminate the effects of error in the square, rotate it 180° around its own vertical reference surface and re-run the test. Subtract one set of readings from the other to arrive at the machine's squareness error.

To measure the relationship of two horizontal axes, place the square on its side. Roughly align one leg of the square with one axis by taking measurements at the extreme ends of that leg, and adjusting the square's position until both readings are at zero. Then traverse and measure along each axis in turn at appropriate intervals, calculate best-fit lines for both axes, and add or subtract one slope from the other, as appropriate.

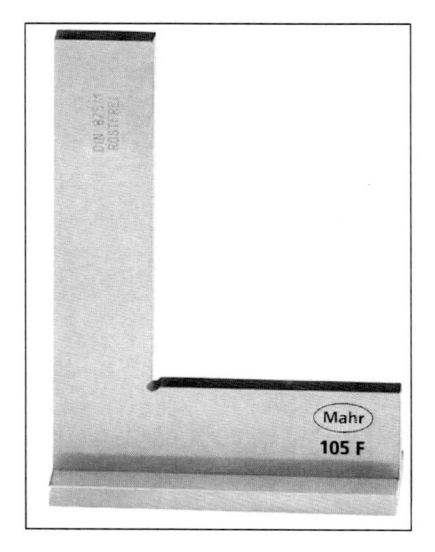

Machine tools can be evaluated for squareness using precision "artifacts" like this mechanical square, in combination with a test indicator or an electronic gage head.

The process is similar when using a mechanical straightedge mounted rigidly on an indexing table, which is itself mounted to the machine table. One axis is measured along the surface of the straightedge, then the indexing table is rotated 90° to measure the other axis. An optical straightedge may be substituted for the mechanical one on the indexing table, with a measurement laser and a plane mirror interferometer used instead of the gage head as the traversing "contact." As the interferometer traverses the optical straightedge, it measures the air gap between them. The 90° reference in this case is still the indexing table, not the optics. Either way, calibrated error in the indexing table should be figured into the squareness calculation.

For machines with more than three feet of travel, a laser interferometer used in combination with an optical square and a straightness reflector is the most accurate method—and probably the only practical one. This method relies upon no mechanical artifact for a reference: the straightness reflector (not the optical square) serves as the reference, while other elements of the optical system must be reoriented between the two straightness measurements. We'll go into more detail on laser measurements in the future.

No matter which method is selected, squareness between axes is a condition that must be examined during machine building and rebuilding, during installation, after a crash (always!), and as part of a periodic evaluation program designed to ensure the geometric accuracy of machine tools.

Evaluating Machine Tools With Lasers

We've looked at several of the measuring tools used for machine tool evaluation, including telescoping ball bars, electronic levels, and mechanical "artifacts" such as precision straightedges and squares. While all of these play important roles in assessing a machine tool's positioning or orientation accuracy, the laser interferometer is the premier evaluation tool. It's not the simplest to use, nor is it inexpensive, but laser measurement systems provide greater accuracy, over a longer range, for a greater variety of measurements, than any other evaluation tool.

Lasers produce coherent light—that is, all the photons move in the same direction in measurable waves—and this is the key to their use as measuring devices. Using relatively simple optics, a laser beam can be split into two halves, each having half the amplitude of the original beam. If these two beams are aimed at the same point, and they travel the same distance, they arrive in phase, recombining as a single beam with the same amplitude as the original beam. On the other hand, if one of the beams takes a slightly longer path, the beams will be out of phase when they arrive, and the amplitude of the recombined beam will be lower, because of "interference." If one path is longer by exactly $1/2$ the wavelength of the light, the two cancel each other out, and the resulting amplitude is zero. Changes in amplitude can be measured with a photo detector, such as a photodiode.

In a typical setup to measure the linear positioning accuracy of a stationary-spindle, moving-table machine tool, the laser is mounted on a tripod beside the machine, aligned with one of the linear axes of motion. An optical beam splitter is mounted on the machine's spindle, and a reflector is mounted on the machine's table, at the end furthest from the laser. The laser is directed into the beam splitter, and one of the two resulting beams is directed by an attached reflector back through the beam splitter to the laser sensor. This becomes the fixed, or reference leg of the measurement path.

Meanwhile, the other beam—the measuring leg—passes through the beam splitter, proceeds to the second reflector at the far end of the table, and is returned through the splitter and recombined with the reference leg. This is where the "laser interferometry" begins: often, the beam splitter/reflector combination is referred to as an interferometer.

As the machine's table is repositioned, changes in the distance between the

This is the basic layout for measuring linear positioning accuracy on a stationary-spindle, moving-table machine tool. Note that the laser is located next to, not on, the machine.

reflector and the interferometer will change the amplitude of the recombined laser signal. These changes in amplitude are converted to "counts," which represent the amount of displacement seen between the two sets of optics. This very accurate measurement is compared against what the machine's controller says is the table position, for an assessment of the linear positioning accuracy.

Several tests are normally run in both directions, to assess repeatability, and backlash or reversal error. Modified setups and different optics can be used to evaluate other machine characteristics, including: straightness of travel, squareness, flatness, pitch, and yaw.

Since the system is measuring changes on the scale of the wavelength of light, positional changes can be measured to 0.001 µm (or 0.1 microinch). This is far higher resolution than can be obtained with any other evaluation method. (Of course, environmental factors must be carefully considered and corrected to achieve accuracies at these levels.) Furthermore, lasers can be used to measure machine axes over 120 ft.—again, something no other instrument is capable of doing.

Any shop hoping to consistently meet specifications for tight machining tolerances should seriously consider laser-based machine evaluation. For those who wish to avoid the expense and training involved, there are service companies that can bring the equipment to you, perform the tests, and present you with the results—often with recommendations for corrective actions to improve machine accuracy.

Let's Level

The electronic level is a tool whose value is often unappreciated. It's as easy as using a spirit level. You just have to make sure the surface you are measuring is clean. The issue is not how to use them—it is where.

The electronic level has two general applications. The first is to show deviations from the horizontal by measuring the angular relationship between a surface and the earth's gravity. This function is used to level a machine and to check components for straightness. The second application is to compare the orientation of surfaces. Both applications can be extremely valuable in the machine shop.

The advantage of an electronic level over a spirit level is a matter of resolution. E-levels can resolve to 0.5 microinches. Auto-collimators and laser calibration systems can perform the same tasks, but E-levels are much faster, less expensive and do not have line-of-sight requirements.

You can use the level to check any surface

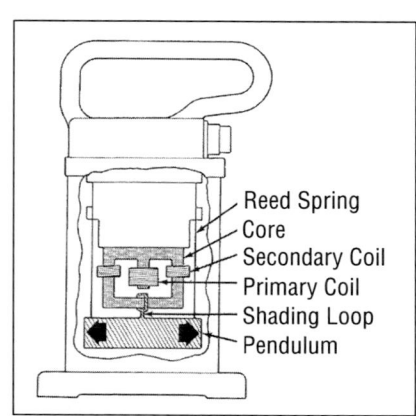

Components of an electronic level's sensing head, which delivers a signal displayed on an amplifier in seconds of arc. The system offers greater resolution than spirit levels and is faster, less expensive and easier to use than auto-collimators or laser calibration systems.

that is critical to machine function. Such level checks are especially useful when setting up a new machine.

Straightness checks are valuable in acceptance testing of new machines or surface plates. You can also use them to optimize the accuracy of an existing machine by avoiding sectors of the ways that are not straight. You may be able to obtain better performance than the manufacturer guaranteed by operating in the "sweet spots" on the ways.

Two sensing heads can be connected to a single amplifier and set for opposite responses to a common motion. A tilt to one side will measure a positive unit on one sensor and a negative unit on the other, canceling each other out. This setup can be used to sense differences in levelness between two surfaces, or between areas on a single surface. Because only differentials are sensed, the object being measured does not have to be level.

You can check the flatness of a surface plate or machine bed by keeping one sensor stationary and moving the other to produce a series of straightness plots. The plots can help you isolate the best sections of the surface, or indicate whether or not resurfacing of the plate is required.

The electronic level is so useful, and so easy to use, it is a crime that it is not more popular.

Using Differential Levels

A differential level consists of two electronic level sensors connected to a single gaging amplifier or readout. In differential mode, the levels are set up so that one outputs a positive signal, and the other outputs a negative signal. Tilting both levels at the same angle causes the signals to cancel each other and the readout to show zero. This is useful for avoiding the effects of local vibration when calibrating a machine tool. In use, one level, which is defined as the reference, is set at zero and remains stationary throughout a gaging trial. Measurements are taken from the second, active, level at various locations. The readout indicates measurements in units of arc-sec., plus or minus, relative to the first level.

Absolute levelness—i.e., levelness relative to gravity—can be measured with a single electronic level. This is useful when setting up a machine tool for the first time, but it is not critical to machine performance. As the ANSI/ASME B5.54-1992 standard on machine calibration states, "Those who doubt this should note that machine tools work perfectly well aboard a ship." It is the levelness of various components on a machine tool relative to each other—in other words, parallelism and squareness—that affects tool point accuracy.

Deviations from parallelism and squareness can generate significant machining errors. Consider: an angular deviation of just one arc-sec. produces linear displacement of five microinches for every inch of travel. On a machine axis with 60" of travel, that adds up to .0003" of error. Then consider that 1 arc-sec. is just about ideal. Most unqualified machine tools exhibit 10-15 arc-sec. of error at various points along their travel.

Angular error can't be entirely eliminated, so it is important to measure it, in order to compensate for it. That's where differential levels shine: in measuring pitch and roll along horizontal axes, and pitch and yaw in vertical axes. (Horizontal-axis yaw and vertical-axis roll involve no deviations in levelness and must be measured by other means.) According to B5.54, angular deviations can also be measured with a laser interferometer or an autocollimator, but level systems are much cheaper, easier to set up and use, and capable of resolution and repeat accuracy as high as 0.1 arc-sec.

To measure roll along a horizontal axis (assuming a vertical-spindle machine), the active level is placed on the worktable, square to the axis being measured, while the reference level is held by the spindle, using a simple fixture called a spindle block, as shown in the figure. The table is traversed along the axis, with measurements taken at appropriate intervals: for 60" of travel, 10 or 12 evenly spaced data points should suffice. To measure pitch on the same axis, the levels are simply turned 90° and the process repeated. To measure pitch and yaw in the vertical axis, the level on the table would be defined as the reference, and the vertical axis traversed.

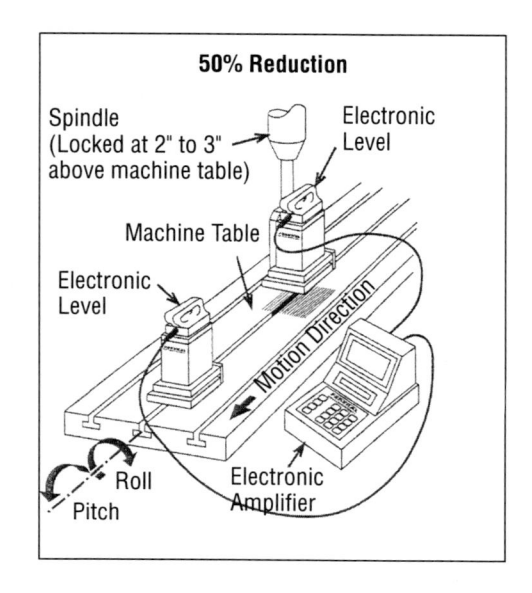

For final qualification, B5.54 calls for a high degree of repetition. Tests are run with the table traversed in both directions, after which the levels are turned 180° and the table is again traversed in both directions. Because of the ease of using electronic levels, this redundant testing can be accomplished quite quickly.

Based on the results, the machine is adjusted, and the tests rerun. The remaining angular deviation can be converted into linear error for any point within the work cube, using the formula: one arc-sec. = five microinches/inch. From this data, a tool point error chart for each axis can be created on a computer. The operator enters the appropriate figure under "work surface to tool point distance" for the cut, and the computer generates a new chart showing the correction required for any location along the axis.

■ ■ ■

Use A Straightedge To Assess Machine Tool Accuracy

Machine tool evaluation can be used to detect problems such as lack of straightness in the travel of a machine's carriage. This can be extremely helpful in explaining why the groove you are trying to mill won't come out straight and square, and it can even suggest corrective actions to solve these problems.

But while it is growing in popularity in the most quality-conscious shops, machine evaluation is not exactly common practice. At its highest levels, evaluation can be a time consuming procedure, with a steep learning curve and a high-dollar barrier to entry.

But don't despair of obtaining the benefits of evaluation if you are short on time or cash. Many useful tests can be performed with simpler, less expensive equipment—some of which you may already own. It is possible, for example, to measure the straightness of a machine tool's horizontal axes of motion using just a straightedge and an electronic gage head and amplifier—or even a mechanical test indicator.

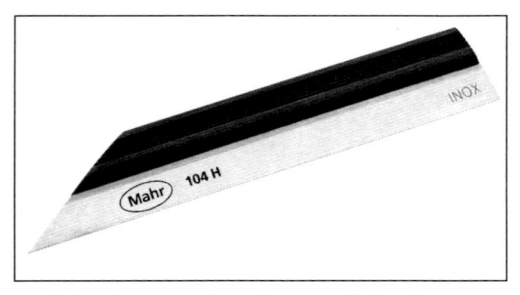

In this context, a straightedge is a certified "artifact" that is traceable to a known standard and accurate to a high degree. A typical glass or steel straightedge with a straightness specification of 10 microinches over 24 inches can cost over $3,000. A granite straightedge costs just $500 or so, but its accuracy is also less: about 50 microinches. Glass straightedges cost less than steel, but because of their fragility, they are usually confined to lab environments.

Although refinements are possible, the basic principle of straightedge-based machine evaluation is straightforward. The straightedge is placed on the carriage, parallel to the axis that is being tested, while a gage is mounted on a stationary part of the machine tool, with the sensitive contact positioned against the artifact. Assuming that we are traversing the table along the X axis, we can measure straightness in the Y or the Z direction, depending upon the orientation of the gage head and the straightedge itself.

Using an electronic gage head, amplifier, and chart recorder or computer, one can produce a continuous trace of straightness. If one is using a mechanical, lever-type test indicator, discrete readings can be taken at intervals of one inch or two inches.

When testing the Z (vertical) straightness of a horizontal axis, the straightedge should not be placed directly on the machine's carriage, because a lack of flatness in the carriage could telegraph itself right through the artifact. It is, therefore, important that the straightedge be supported properly by two precision blocks or parallels. The proper placement of these supports at the "points of least deflection" (length × .554) is critical to achieving maximum accuracy.

Regardless of a straightedge's certified level of accuracy, it can be improved upon using a "reversal" technique, which lets the user measure straightness errors in the artifact itself to microinch accuracy. The procedure is similar to that described above, except that, after the first run, the straightedge is flipped 180 degrees around its long axis, and the gage head is also repositioned on the opposite side of the machine, so that it contacts the same edge of the artifact. The gage head's direction of motion is reversed, so that a "bump" on the straightedge should read as a positive numeral on both trials. The test is re-run, and the results of one trace are subtracted from the other. Because the gage head was reversed, carriage errors cancel each other out, so any remaining deviation from zero reflects error in the straightedge. (This assumes that carriage errors are repeatable.) The results can be used as correction factors for all future uses of the straightedge.

Whether to use a straightedge or some other method depends on the level of accuracy required, the equipment and financial resources available, and the machine tool. A two-foot straightedge won't work very well on a machine with ten feet of travel. On the other hand, it's hard to fit all the necessary laser optics on a machine with a work envelope of one cubic foot.

A word of caution: In the interests of space, the above descriptions have been somewhat over-simplified, but the general idea remains valid: simple tools, such as straightedges, can serve a valuable function in optimizing machine tool accuracy. Check out their capabilities as a first step on the road to a more complete program of evaluation.

Section I
Gage Blocks

Getting Ready For The Microinch Revolution

Many of us remember when the typical tolerance for machined parts shifted from thousandths to "tenths," and we had to adjust both our production techniques and our QC measurement methods. As ever more technical products make the need for precision even more profound, a similar shift is now under way, toward millionth-measurements. This time, our gaging methods will have to change radically.

Previously, we focused much of our attention on the gage itself: as long as the instrument was designed to the required degree of accuracy and maintained properly, we could usually get by, even at the "tenths" level. Now that we're trying to measure tolerances of 50, 30, or even 20 millionths, we must shift our focus to the measurement process and the environment in which it takes place. Where temperature and cleanliness were formerly somewhat abstract issues, they now become essential concerns.

According to the old gage-maker's 10:1 rule of thumb, when measuring tolerances of 30 millionths, the gage should demonstrate repeat accuracy of 3 millionths. But consider this: a difference of 1°F between the part, the master, and/or the gage can introduce an error of three millionths. In other words, if we don't control thermal influences, we give up all hope of repeatability.

Microinch gaging therefore must be performed in a controlled environment—a special room that is thermally insulated from the shop floor. Temperature should be kept as close to 68° as possible, and changes must not exceed 2° per hour. When a part comes in from the shop, it should sit for several hours on a heat sink (a large steel plate), to bring it into equilibrium with the master and the gage before being measured. Even with all of these precautions, the gage should be mastered frequently.

The gage should be protected from the operator's body heat, and his breath, by a clear plastic shield or full enclosure. The operator should not touch the parts or masters directly: insulated tweezers, gloves, or similar measures should be employed.

Elaborate measures are also required to combat the problem of contamination. Relative humidity in the room should be kept below 50 percent to inhibit the formation of rust. Parts must be thoroughly cleaned of dirt and even thin oil films prior to gaging. The choice of cleaning solvent will vary with the application and may require some trial and error to ensure that the solvent itself doesn't leave a film. It will be necessary to regularly clean the entire gaging area, plus the gage and masters, to remove dust, skin oils, etc. Even the choice of furniture upholstery and the clothing worn by operators must be considered: natural fibers shed more dust than synthetics. The room should have an air lock, and unqualified personnel should be prohibited from entering. If there is a computer printer in the room, it should be in an enclosure, and single-

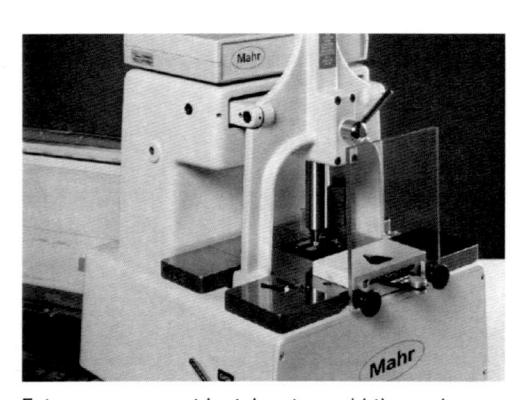

Extreme care must be taken to avoid thermal influences when measuring to microinches. A clear plastic shield blocks the operator's body heat, and insulated tweezers are used to handle the gage block.

sheet paper feeding should be used: paper dust may release into the air when tearing continuous forms along their perforations.

At microinch tolerances, dimensions change so readily that it may be more practical to check part relationships (i.e., clearance) than absolute dimensions. "Match gaging," which is sometimes associated with imprecise process control, can be the best and easiest way to ensure that parts will fit together with the required clearance.

Surface finish and part geometry become critical parameters at the microinch level, and for any degree of repeatability to be possible, it is necessary to use witness marks, or some other method to ensure that a part is always measured at the same location. The whole subject of mastering, calibrating, and certifying a gage to millionths is important and will be examined in the next column.

Even with sophisticated gages that are fully capable of the task, measuring to millionths remains a challenge. It requires thorough planning, careful selection of conscientious personnel, and significant investments in training and facilities, as well as a good understanding of all the variables that can affect microinch dimensions. It may be tempting to just buy the new gage and give it a go, but I can guarantee you'll spend more time figuring out your problems and ways to fix them than you would by doing it right in the first place.

The Mechanics Of Millionth Measurements

In the previous column we looked at some of the challenges inherent in measuring parts to microinch tolerances. We discussed the need for a climate-controlled environment and absolute cleanliness, but we've only scratched the surface (so to speak). Special attention must also be paid to the selection of the gage and readout, and to mastering.

If you're checking parts for tolerances of 10 microinches, or checking gage blocks, rings and discs to millionths, you'll need resolution (i.e., minimum grad value) of 1 microinch, or maybe even 0.1 microinch, on the gage readout. But beware of excessive magnification. Some gage manufacturers create the appearance of microinch accuracy by supplying an electronic amplifier, with units reading in millionths, on a garden-variety gage. Because the average shop-floor gage is mechanically repeatable to only 50 millionths or so, what you really get is a highly magnified look at the gage's repeatability error. The resolution of the readout should accurately reflect the precision (i.e., repeatability) of the gage itself. This can be checked by repeatedly measuring a part under controlled conditions. If the reading varies by .000005" or more between trials, the gage may be incapable of handling microinch inspection duties.

When measuring high-accuracy IDs, gages typically contact the workpiece with up to four ounces of gaging force. At microinch tolerances, this can produce measurable deflection of the jaws. That deflection may be unavoidable, so repeatability demands that it remain constant from one trial to the next. A frictionless mechanism, such as a reed spring arrangement, is therefore essential to maintain the gaging force constant to within $^1/_2$ gm.

Normally, we rely on a master to assure the accuracy of a gage. But if we're using the gage to measure a master, we have to go one step further and refer to certified gage blocks to master the gage. This is relatively straightforward for outside measurements, but more complex for inside measurements.

Fig. 1

There are two accepted methods: In the first, a single gage block or a stack of blocks is used to set a pair of caliper blocks at a precise distance. The blocks are held together with clamping rods, with the caliper blocks extending over the gage blocks on both sides. (See Fig. 1.) The gage is mastered to the distance between the caliper blocks, checking at both ends for parallelism error. Ball feet may be snapped onto the block assembly: this makes it easier to move around on the table using insulated forceps or other tools, and helps to minimize the transfer of heat from the operator's hands.

In the second method, two or more blocks are wrung to a true square, with enough overhang on the topmost block to fit over the gage contacts. A second set of blocks is wrung against an adjacent edge of the square to check for squaring error. (See Fig. 2.) This setup is less subject to parallelism error than the first method. In both methods, the operator should wait at least three hours after assembling the blocks, to give temperatures a chance to normalize.

When measuring master rings, it is necessary to account for geometry errors, as well as surface finish, scratches, waviness, and other microinch imperfections. Measure Class XX masters (tolerance = .00002" for sizes between .029" and .825") at a depth of $1/16$" in from both ends, and in the middle of the ring. This avoids the bellmouth conditions likely to exist near the ends, and detects most barrel-shape, hourglass, and tapered conditions.

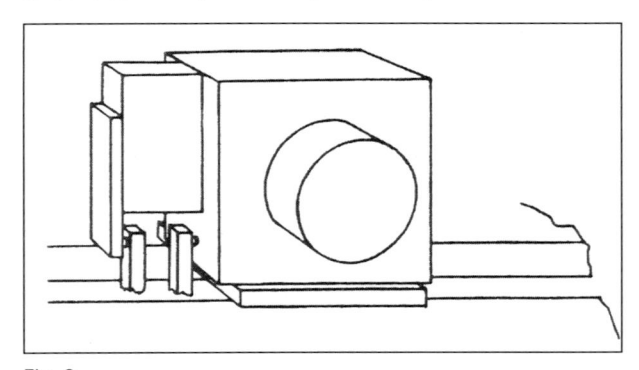

Fig. 2

For Class XXX masters (tolerance = .00001" for sizes between .029" and .825"), the drill is even more rigorous. Confine all measurements to a $1/4$" tall band in the center of the ring, and take a total of six measurements: at the top, middle, and bottom of the band, in both north/south, and east/west orientations.

The point here is not to find and discard masters that show variation of a millionth or more. When working at the microinch level, a certain degree of uncertainty is inevitable. The objective is to minimize the uncertainty that the master contributes to the overall measurement process.

Don't Wring Your Hands, Wring Your Gage Blocks

You need a master for something you have to measure, so you reach for a set of gage blocks and begin creating the dimension. But how are you going about it? Some machinists, with experience using gage blocks, will begin creating the desired dimension by starting with the largest block first and grabbing blocks as they go. Then they get stuck and start wringing their hands.

However, there's more to combining gage blocks than just "stacking" blocks. When done properly, you can create any dimension from zero to four inches, to the nearest tenth of a thousandth, never using more than four blocks or never using more than one from each series of blocks. In addition, you are assured of minimizing wringing error and randomly spreading the usage among all the blocks, thus reducing wear on the blocks.

Let's say the dimension we need is 1.3248 inches. We've got four series of blocks, as shown in Table 1. Now, to create 1.3248 inches, rather than saying, "I need a one inch block," we start at the right side of the dimension and work backwards, with the goal of creating zeros all the way across.

Table I — Four Series of Blocks				
Series	No. of Blocks	From	Through	Increments of
1st (Tenths)	9	0.1001"	0.1009"	0.0001"
2nd (Thousandths)	49	0.101"	0.149"	0.001"
3rd (Fifty Thousandths)	19	0.050"	0.950"	0.050"
4th (Inch)	4	1.000"	4.000"	1.000"

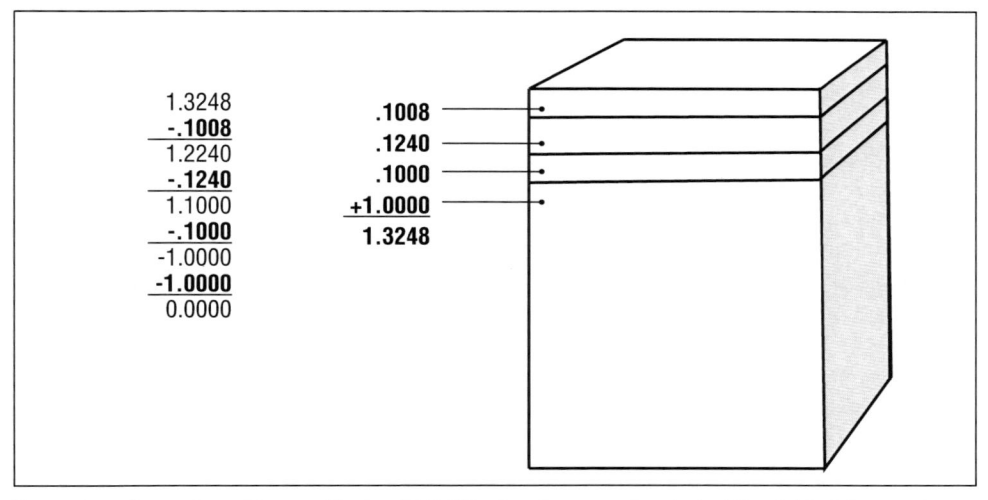

```
 1.3248          .1008
-.1008           .1240
 1.2240          .1000
-.1240         +1.0000
 1.1000          1.3248
-.1000
-1.0000
-1.0000
 0.0000
```

To create a dimension with gage blocks, 1.3248 inches for example, start at the right of the number and work backwards, with the goal of creating zeros all the way across. When blocks are properly combined, any dimension from zero to four inches can be created to the nearest tenth of a thousandth, never using more than four blocks, or more than one from each series of blocks.

To get rid of the 8, there's only one series with an 8 in the fourth place: the Tenths. So we select the 0.1008-inch block. Subtract from 1.3248 inches, we're left with 1.2240 inches.

Now we want to get rid of a 4 in the third place. We have more than one in the Thousandths Series blocks with a 4 in the third place: 0.104 inch, 0.114 inch, 0.124 inch, 0.134 inch and 0.144 inch. To get rid of the 4, as well as the 2 in the second place, we select the 0.124-inch block. Subtract from 1.2240 inches and we are left with 1.1000 inches. See how quickly this system gives us round numbers.

We have a 0.1000 block in the Fifty Thousandths Series, which leaves us with 1.0000 when subtracted from our number. Finally, to get rid of the one-inch number, we have the one-inch block in the Inch Series. So, the math for 1.3248 inches looks like that shown in the example above.

By working from right to left, we have created zeros all the way across. We have not used more than four blocks, and only used one block from each series.

We could create the same dimension using many more blocks, but there are only three "wrings" in a stack of four blocks. There could easily be wringing error of four-millionths per wring. The more blocks we use, the greater the wringing error, and the more the blocks are exposed to the shop environment, where they may get dirty, scratched or lost. We need to combine gage blocks properly for the same reasons we need to give them lots of care.

■ ■ ■

Clean It Up And Wring It Dry

There's probably nothing in your shop that is made to the same high level of precision as your gage blocks, with their manufactured tolerances closer than a few millionths. The Hubble Space Telescope gets mirror envy when it thinks about the smoothness and accuracy of a common gage block. It requires just a few, but nonetheless important, care and maintenance measures to retain that extraordinary precision for years.

Cleanliness is critical; it is Rule Number One. Never wring, or use in any other way, gage blocks that have been exposed to chips, dust or dirt. Blocks that have been exposed to cutting fluids must be cleaned, without fail, prior to wringing, or the metal particles held in suspension will surely wear the blocks' surfaces.

To clean blocks, use filtered kerosene, a commercial gage block cleaner or some other high-grade solvent that doesn't leave a residue. Wipe them dry with a lint-free tissue. Even if you are taking "clean" blocks from their storage box, clean them again. They've probably picked up lint or dust while in the case.

Wring blocks together "dry." Rubbing them on your palms or wrists will deposit oils that may assist in bonding the blocks during a wring, but it may also transfer dirt and moisture that can damage the surfaces. If you can't get a good wring dry, use a commercial wringing solution, available at most trade stores.

Don't allow gage blocks to remain wrung together for long periods, because they can become permanently fused to each other. If you use the same setup day after day, make sure you separate and then wring them daily.

Frequently inspect the blocks for nicks, scratches or burrs, and repair or replace any damaged blocks before using them. One advantage of combining blocks by the method described in the previous column (by zeroing the dimensions in order, from the smallest decimal place to the largest) is that wear is randomly spread among the entire series of blocks. But if even one of those blocks is damaged, you will end up randomly transferring scratches among your entire collection.

When you are through, make sure you always clean the blocks before putting them away in their case, and coat them with non-corrosive oil, grease, or a commercial preservative. If you don't coat them, they will rust, even in the box. Just think what corrosion will do to the surface of blocks when you are wringing them.

Regard the case as part of the working system of gage blocks. Return blocks to the proper slot in their case as soon as possible. Beyond protecting them from dust, the case ensures that the blocks don't get tossed haphazardly in a random box, where they can easily damage one another. The labels on the case are much easier to read than the numbers etched on the blocks themselves, so you will spend less time looking for the right block. And you will be able to see immediately from the empty slots what blocks are in use or missing.

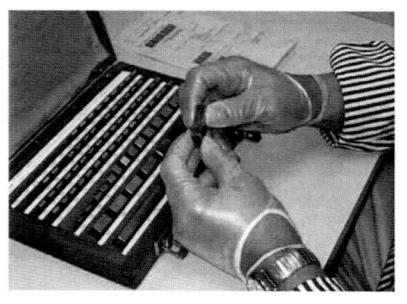

Keep gage blocks in their storage case, and clean them not only before they're put away, but also when they're taken out. Use filtered kerosene, a commercial gage block cleaner, or other high-grade solvent; wipe the blocks dry with a lint-free tissue; and coat them with a preservative before putting them away.

Keep the storage case scrupulously clean, both inside and out. This will serve to remind you and other users that the gage blocks are not just hunks of metal. We are talking precision instruments here, and they must be treated as such to retain that precision.

Finally, don't loan out individual blocks. The minute a set is broken up, all of the above points are likely to go for naught. You can no longer control the conditions under which those blocks may be used, and they may return to haunt you by transferring dirt, corrosion, or scratches to your other blocks. That is, of course, if they return at all.

■ ■ ■

Working With Your Working Gage Blocks

Working gage blocks are the ones used when the rubber meets the road in today's shop environment. Their use is as varied as the number of gage blocks found in a large set. The working blocks have an intermediate grade, and are often used in the inspection or calibration lab, but may also be found out on the shop floor. We are not talking about the Master Gage Block set, which is used as the corporate standard, or the shop floor set which may sometimes level out an old table. As part tolerances become tighter and the resolution of comparative gaging higher, the use of working gage block sets for tool making and part inspection is very widespread.

Of course, the most common use of the gage block is to provide a reference for direct measurement of distances between parallel surfaces, such as widths of grooves, slots, etc. The blocks can be used as a go/no go gage or as a means of setting up a comparative measurement. The ability to stack the gage block set to any length is what makes the gage invaluable as a tool in the inspection lab.

A second important use of the working block is for checking the performance of the shop hand tools and gaging equipment. For example, micrometers and other hand measuring instruments can be checked for linearity (degradation of the micrometer thread). Using gage blocks of different sizes that are stepped, usually in equal increments over the measuring range of the measuring tool, provides a means for checking the performance of the tool. The gage blocks are also useful for setting a reference point on measuring tools where the measuring capacity of the gage is larger than the measuring range of the gage itself. For example, a 3" gage block would be used to set the 3" reference point on a 3" - 4" micrometer. With these measuring tools the spindle does not touch the fixed contact, so a gage block provides a precision method of setting a starting point.

Limit gage sets can be created with two sets of gage blocks or gage block stacks set to the high and low limits of the tolerance. These sets would consist of gage block stacks with added contact elements or end jaws all combined together to the limits specified. They provide an easy-to-assemble and reliable means of producing temporary gaging for an unexpected application. However, in actual use it is much more economical over the long run to have a fixed gage made up for these applications, as the constant assembly and disassembly is time consuming and subject to stacking error.

In the manufacturing inspection area a frequent use of these blocks is to set up for comparison or transfer measurements during surface plate work. Using a single block or stack for a height reference is the common approach. By adding another gage block or endjaw, which over-hangs the block, an inside dimension can be established. A base, holding rods, and endjaws, are often found in a gage block accessory kit. They allow the assembly area to perform medium accuracy comparisons with surface plate work. Finally, the use of gage blocks and a sine plate on the surface plate create a very precise angle setup. The double requirement of accurate length and perpendicularity to the base is a perfect use of the working gage block. Using fairly straightforward trig, the right angle reference set up by the blocks and the known length of the sine plate makes for easy angle setup. Just as it's necessary for you to get ready to go to work, gage blocks need to get ready to work also. Since they can't do it by themselves, it's necessary to help them out. Preparation includes making sure they are recently certified, checked for nicks and burrs, and stabilized for the temperature of the work area.

With some care and careful handling, working gage blocks will make a significant contribution to the productivity of a manufacturer.

Gage Block Verification

I got a note the other day from a QC manager wanting to know if it would be acceptable for him to verify his gage blocks in-house. "You always say the gage-maker's rule of thumb states that any gage used to verify another gage must resolve to ten times its accuracy," the gentleman wrote. "We've got a perfectly good bench gage here that resolves to millionths of an inch: can we use that?"

The answer, like so many in metrology, is 'it depends.' The rule of thumb does say 10X accuracy, but there are also *fingers*. Take a look at the following measurement requirements for verifying gage blocks, and then you decide.

Gage blocks are used throughout the production process, but those used in the inspection area, tool room, or on the production floor are especially subject to wear. Even age can affect the size of a gage block. And with tolerances growing ever tighter, regular verification of gage blocks is now more important than ever.

The most widely used method today is still the comparison of a test gage block to a reference or master gage block of the same size, and then recording and documenting the variation between the two. There are two methods of comparison—comparison by interferometry or mechanical contact methods. The mechanical method is by far the most common in industry today, and it is done on a special class of comparator specifically designed for this purpose.

While these instruments may look and operate very much like typical precision comparators, there are many additional requirements needed to achieve millionth-level measurement gage block verification. This is not to say that other comparators cannot measure to millionths of an inch, but gage block comparators are designed to take these additional requirements into account.

For a 'regular' precision comparator to achieve reliable millionth-level measurements, it must meet some fairly stiff criteria:

- A large and massive base—one that does not change in size quickly with slight temperature changes.
- A rigid base—at this level, bases, posts or arms that might deflect can create errors even bigger than the intended measurement.
- A high-resolution amplifier is needed to discriminate to the tolerances required. Resolutions need to be one millionth of an inch, or even less, and it must be stable and not drift.
- A highly linear gage head that has excellent repeat characteristics.

But for a gage block measuring system, the following characteristics are also needed:

- Two measuring contacts, both contacting the block differentially. This eliminates form error from the block measurement. Because the two surfaces of a gage block are not perfectly flat and parallel, and because any reference surface is also not perfect, size measurements will depend on exactly where the measurement is taken. Contacting at two specific points, exactly opposite one another on the gage block, is the way to get good, repeatable measurements that can be reproduced from year to year.
- Contact points of known material (diamond) and of specific radius.
- Known and constant gaging pressure over the full measuring range.
- The ability to electronically adjust mechanical zero. Mechanical zeroing would be nearly impossible at this high magnification.
- Retractable contacts to eliminate any possibility of scratching the block or wearing the contacts.
- A platen to support—but not measure—the block, with grooves to help keep the block clean.
- A method for positioning the gage block to specified measuring locations as called out in industry standards. New requirements specify blocks be measured at specific locations.
- A means of quickly setting the gage to another size with as little mechanical adjustment as possible. Since a typical gage block set may consist of 81 blocks, the gage should have the ability to set size rapidly, with as little influence on the gage as possible.

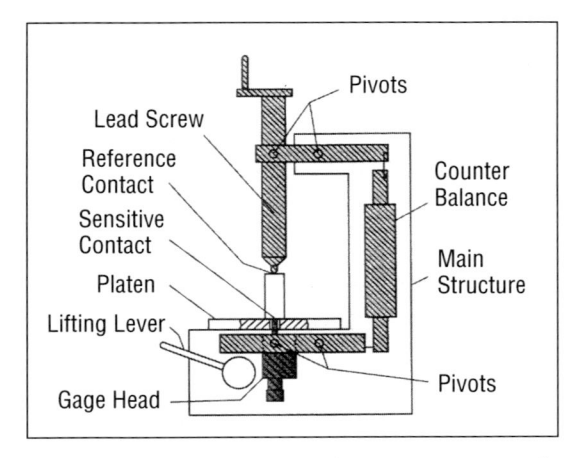

Since gage blocks need to be measured with two contacts, differential gaging is a typical method to accomplish this. Differential gaging uses two high precision LVDT type transducers. These are very linear and repeatable over short ranges, and are very good for these applications. But there is also an alternate method available on some systems that is even better.

In any LVDT, there is some very small error stemming from linearity, repeatability, or slight changes in gaging pressure. Eliminating one gage transducer, while still performing a differential check, eliminates any possible error associated with one of the heads. By mechanically creating a differential measurement with a floating caliper, errors can be cut in half and the overall performance of the gaging system improved.

Finally, while gage block comparators are designed for millionth measurement that takes into account all the physics of good gage block verification, software programs further reduce any sources of error. For example, a gage block program can correct for minute errors resulting from gaging pressure of different gage block materials and changes in temperature.

Thus, while there are many systems like bench gages and horizontal measuring machines capable of measuring in the millionths of an inch, only a gage specifically designed to measure gage blocks provides the ultimate system for both speed and performance.

Two excellent references for our QC manager are (a) THE GAGE BLOCK HANDBOOK, NIST Monograph 180, available from the National Technical Information service, Springfield, VA 22161 and (b) American National Standard B89.1.9-2002, Gage Blocks, available from The American Society of Mechanical Engineers.

New Gage Block Standard

Location Is Everything

One of the few things you can count on in life is that ASME's Dimensional Standards are relatively stable for a long period of time. This is good because change is a hard thing to deal with— especially in the metrology world. The United States Gage Block Standard has not changed much since 1970 when Federal Specification GGG-G-15B was introduced.

However, there have been many changes over the years in the use of gage blocks: in block materials, in the internationalization of manufacturing, and changes resulting from the influence of new measuring principles such as uncertainty and traceability. Thus, in 1990, ASME B89.1.9 started the ball rolling towards updating the standard by introducing more international influence. Finally, in ASME B89.1.9 2002, the Gage Block Standard was revised in an effort to bring it closer to ISO 3650, while at the same time incorporating our desire for inch units, square blocks and some of the grade requirements ingrained in North America.

While most of the changes in the new specification don't affect the everyday user of the blocks, they may influence how gage blocks are identified (or graded), how they may be measured, and how the information is presented. The standard has already affected gage block manufacturers in the way they specify their products.

Actually, ASME B89.1.9 isn't bad reading. Besides the normal terms, definitions and tables, it presents some good information on referencing from the old standards to the new, on the handling of gage blocks, and on typical set configurations. It even lists very nicely the differences between previous standards and the new one. This is done to clear up different terminology and to match the different grading conventions, old and new.

As regards grading, there are a couple important items to point out. The first is that grading is now modeled much closer to the ISO method.

There are also additional categories for calibration grades of blocks, designated as Grade K or 00, that are used as reference standards.

The other noteworthy change is in the actual tolerance of the blocks. Gone are the unbalanced tolerances such as +16, 8µ" as was seen for a grade 3, 2.0" block. Today, that tolerance is +/- 16µ" for the equivalent grade AS-1.

Finally, a change was made to add a specification for length variation of the block. This was previously referred to as parallelism. Today, each grade has a length variation tolerance that is dependent on the grade of the block. What is also important is that this length variation can be as small as 2µ" and the standard specifies where it must be measured. The new requirement is that all blocks be measured at the gage length (which is the center of the measuring face), and at all four corners. Corner measurements must be taken at approximately 0.060" from each of the side faces, and all of the points measured must not only be within the length tolerance for the grade, but also within a specific variation range for the length tolerance. For example, if the grade length tolerance is +/- 8µ" and the tolerance on length variation is 4µ", then all your measurements must fall within the +/- 8µ" and also not vary very far from each other by 4µ".

While this may not seem to be much of a chore, it really increases the burden for the gage block calibration people who do the actual measurements. Previously, they would make four measurements somewhere on the face of a rectangular block, simply by moving the block around with tweezers (precision tongs). Now that the spec requires measuring at an exact location, measurement is much more difficult and time-consuming for the operator.

To ease this situation, special gage block manipulators need to be incorporated into gage block comparators to position the blocks to the exact locations in each corner, every time and very repeatably. Remember this last statement, because with gage blocks we are in the world of microinches. Like real estate, the key to measuring millionths is location, location, location. The better you can position the gage block, the more you increase repeatability and reduce uncertainty.

Eventually all the tools used to calibrate gage blocks will need to incorporate gage block manipulators if calibration facilities are to improve throughput and maintain tight uncertainty levels.

Formulating Sources Of Error In Your Form Measuring System

It is generally understood that the results of precision measurements—such as from a form measuring instrument—are subject to a number of environmental influences, such as shock, vibration, and temperature deviations. What is less well understood, however, is that the form measuring machine itself can also influence the measurement results. For example, worn probes, excess bearing clearance, natural vibrations, and other factors can all degrade the overall accuracy of a measurement. These measuring system-based factors that influence the assessment of form are called 'Measuring Uncertainty.'

Some suppliers of precision parts are required to take the measuring uncertainty into account before delivering their products to their customers. Here's why. Let's say that the specification for radial run-out of a shaft is called out in the tolerance at 3 µm. From documentation we then discover that the uncertainty of the measuring instrument amounts to +/- 1 µm. Thus only shafts with a radial run-out of less than 2 µm can be accepted. Once the measured radial run out values reach or exceed 2 µm, you can no longer exclude the possibility that the inspected work pieces are out of tolerance.

For this reason, it's not too surprising that measurement uncertainties are disclosed for facilities doing measurement standards, such as gage blocks, master rings and discs. But there are also cases where the measuring uncertainties of instruments used for the inspection of products come under scrutiny. Only when you have determined the uncertainty of an inspection system for production parts can you then determine what part of the tolerance band is "left over" for actual production. The drawing tolerances, which are often extremely close, are narrowed even more if the measuring uncertainty is too high. The upshot of this is that imprecise measuring devices increase production effort and, therefore, cost.

The present internationally approved standard for the determination of measuring uncertainty is the GUM method (Guide to the expression of Uncertainty in Measurement). The first procedural step in determining uncertainty is the determination of all the influence quantities. The complexity of a form measurement becomes quite clear during this very first step. Here's a partial list of influencing factors that may cause measurement errors:

1. Examples of environmental influences: Temperature (fluctuations), radiant heat (e.g., from the operator or the lighting), air refraction/gradients (influencing optical based systems, including lasers), humidity, vibrations and shocks.

2. Examples of influences caused by the part being measured: Fixturing method, alignment, distortion through measuring force an dead weight, size and

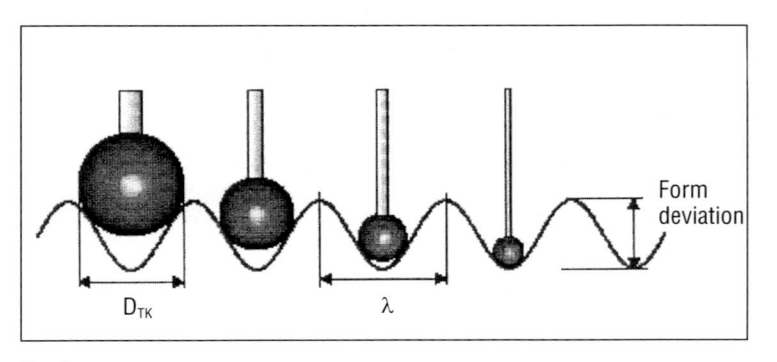

Fig. 1

type of the gaging and datum surface(s), roughness of the gaging surface, undetected form errors of the gaging surface, form errors of the datum surface(s).

3. Influences caused by the operator: Misinterpretation of the drawing specifications; excessive clamping force when fixing the part under measurement; selection of wrong probes (Fig. 1); selection of wrong parameters (e.g., wrong profile filter, excessive measuring speed); programming errors and computation errors; heat radiation; shocks. For this reason, it is important that operators have comprehensive training in metrology, and in the correct adjustment and operation of their measuring instruments.

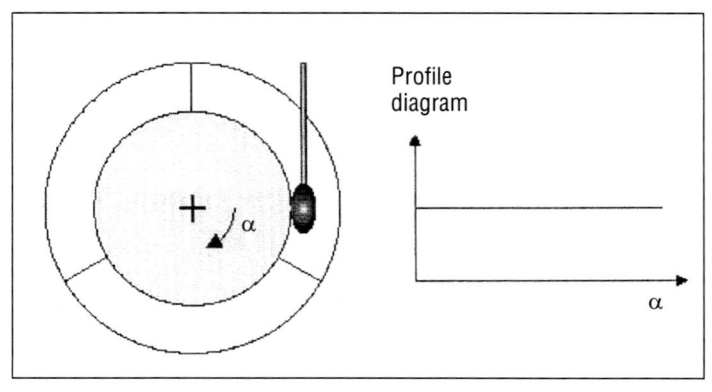

Fig. 2

4. Examples of influence caused by the form measuring instrument: Deviations of the measuring axes, irregular movement during measurement, errors in the electronic indicating and control system (e.g., rounding errors), software errors.

The choice of measuring instrument has considerable influence on the number of factors that determine the measuring uncertainty. For example, compare a sample roundness measurement on a form measuring instrument with a rotating measuring axis, and a roundness measurement on a 3D coordinate measuring instrument. Using an instrument with a rotating measuring axis, as in Fig. 2, shows that the influence of stylus ball form errors on the measuring result is negligible. Fig. 3, however, shows that this does not apply to 3D coordinate measuring instrument: in this case, the stylus ball deviations have to be thoroughly calibrated before measurement. For this reason, the degree of uncertainty for roundness measurements on a 3D coordinate measuring instrument depends in large part on the uncertainty of probe calibration (into which, in turn, the measuring uncertainty of the calibration standard enters, among other factors).

All these factors, which may significantly influence the uncertainty of form measurements, have to be estimated on the basis of concrete data for the expected value and the standard deviation.

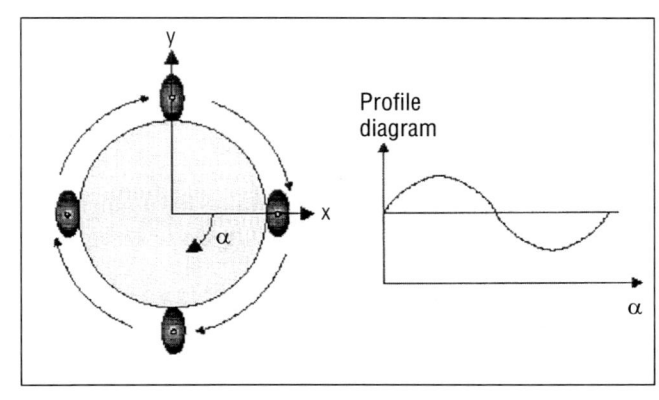

Fig. 3

Section J

Geometry

The Shape Of Things To Come

What do you do when you're doing everything right and it's still coming out all wrong? We had a case like this not too long ago, where a shop was making a spindle and bore assembly for a high precision application. The owner complained that while he was machining well within his specified -.0002" tolerance on the shaft, his parts were either causing excessive bearing loads, or worse, not fitting in the bores at all. "How can they be wrong," he said, "when everything measures right?"

What this fellow didn't realize was that there are more things that can go wrong with a part than dimension. Just as we cannot make parts perfectly to size, neither can we make them perfectly round or perfectly smooth. And, as we all move towards tighter and tighter tolerance machining, irregularities in shape and finish will have a greater and greater affect on our ability to make parts. This means we're all going to have to understand more about geometry and surface finish.

In the example above, analysis in our lab showed a consistent three-lobed out-of-round condition on the spindles, which was making their effective diameters too large. Three-lobed out-of-round is very common when using centerless grinding, but it wasn't noticed in this case because: 1) the specs didn't call for any geometric analysis on the parts; and 2) the shop was only using a two-point dimensional gage which was incapable of detecting the problem.

Fig. 1 illustrates the relationship between out-of-roundness and effective diameter on a three-lobed part. As you can see, any two-point measurement will yield a consistent diameter, because each lobe is geometrically opposed by a flat area. This measured dimension would fall somewhere between the inner and outer dotted circles. However, the <u>effective</u> diameter, or the amount of space this part would actually require to clear, would be the outer dotted circle, which encompasses all the lobes. In this case, because the tolerance was so tight to begin with, the increase in effective diameter caused by the roundness problem exceeded his total tolerance for the part.

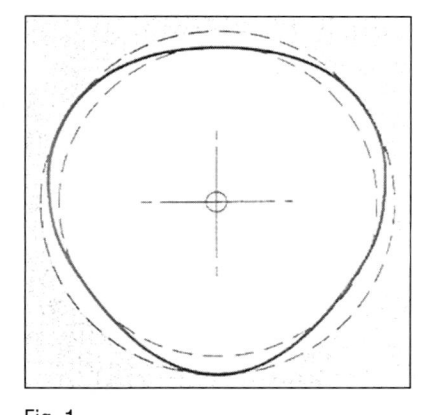

Fig. 1

So did that mean he had to invest in a lot of fancy lab equipment, or buy a new centerless grinder? Fortunately, not. As noted, out-of-round conditions with an odd number of lobes are common with centerless grinding (the greater the number of lobes, the more closely you approach true round), and once understood, are easily compensated for. In this case, a simple V-block fixture was set up with the blocks at 60° to measure the effective diameter, and the grinder set accordingly. Without going through the math involved, other odd-lobed out-of-round conditions can be similarly detected, using V-block fixtures set at other angles (108° for five lobes; 138°40' for seven lobes; and so on).

Unfortunately, in this case (but not for this column!) out-of-roundness was not the only problem. There was also a problem with surface finish, which, while specified, was not really being measured. The specs called for an average roughness (R_a) of no more than 4 µin., but when measured, the parts showed an R_a of between 15 μin. and 25 μin. Since the affect of roughness on overall tolerance is a factor of at least 8, and sometimes as much as 20 (see Fig. 2), the 25 μin. of roughness took up the entire .0002" tolerance range on these parts!

Again, the solution was not costly equipment—surface finish gages are readily and economically available for shop floor use—but an awareness of the problem and an understanding of the basic causes. A simple redressing of the wheel solved the problem here, and allowed our shop owner to resume his normal sleeping pattern at night.

But the lesson is an important one. A recent report by the National Center for the Manufacturing Sciences showed that machining tolerances have decreased by a factor of five within the last decade, and that even tighter tolerances are on the horizon. This means that things like geometry and surface finish are going to play an increasingly important role in machining operations. And we need to understand that role, if we are to continue to produce good parts. That's the shape of things to come.

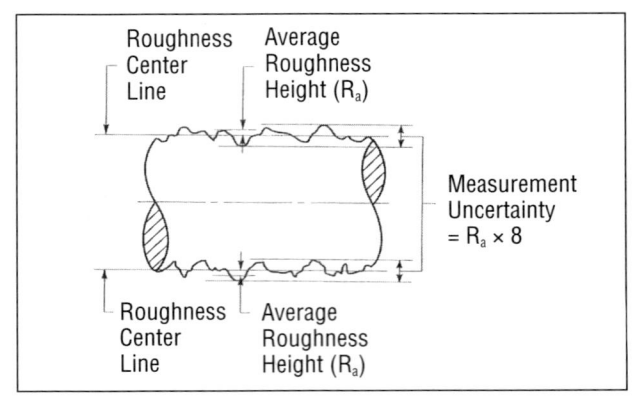

Fig. 2

Roundness Gaging—Approximately

In the last column we discussed the relationship between part geometry (i.e., roundness) and dimensional tolerance. Circular geometry gages, with their precision spindles, are the best—and the standard—method for measuring out-of-roundness. But these can be elaborate pieces of equipment and are usually confined to applications where a very high degree of accuracy is required concerning part geometry.

Most jobs, however, have fairly simple requirements for roundness. While a true roundness measurement requires a complex description of the geometric and dimensional relationships of dozens or hundreds of points on a diameter, most job specifications simply call for parts to be "round within 0.XXXX in. variation in radius." In other words, as long as no point on the radius falls outside of two concentric circles, the actual shape of the surface is secondary.

This being the case, there are ways to approach the problem of roundness measurement that can provide a pragmatic, low-cost alternative to the circular geometry gage. Although these methods rarely give a technically accurate measurement of roundness, they are often close enough to give a good indication of the functional implications of an out-of-round condition. If you understand the nature of your out-of-roundness, it may be possible to qualify the part using conventional equipment such as a micrometer, bore gage, comparator stand or a V-block arrangement.

Section J: Geometry

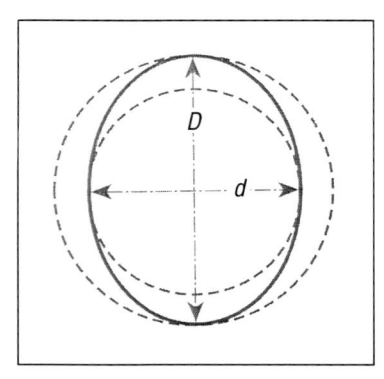

Fig. 1

Understanding the geometry involved is the key. Generally speaking, out-of-roundness is either symmetrical, involving regular or geometrically arranged lobes or points on the part's circumference, or asymmetrical, where the lobing is not regular. Most machining processes create symmetrical lobing, producing either an even, or an odd number of lobes. Even-number lobing is sometimes seen in precision boring operations, caused by a worn or out-of-balance spindle. Odd-number lobing may be caused by a three-jaw chuck (producing a 3-lobed workpiece), or a centerless grinder (which may create a 5-lobed condition). Asymmetrical lobing cannot be measured by the means described here. It is evidenced by irregular travel of an indicator, and is usually indicative of a problem in the tool.

In cases where an even number of lobes is arranged geometrically on the part, each lobe is opposed by one diametrically opposite (see Fig. 1). The piece, therefore, will have major and minor diameters. Knowing this, we can gage the part using simple two-point, or diametrical measurement methods. The difference in the differential measurements will generally be twice the out-of-round value due to the diametral versus radial method of assessment. For example, if our specs call for a part that is "round within 0.0001 in. variation in radius," we can measure using a simple comparator, and reject any part where the Total Indicator Reading (TIR) is larger than 0.0002 in.

Parts with an odd number of lobes pose a slightly more complicated problem. As shown in Fig. 2a, where an odd number of lobes exist, each lobe is diametrically opposed by a flat area. These parts can be measured on a V-block fixture, using the following formulae to establish the included angle of the V-block, and the multiplication factor (illustrated in Fig. 2b).

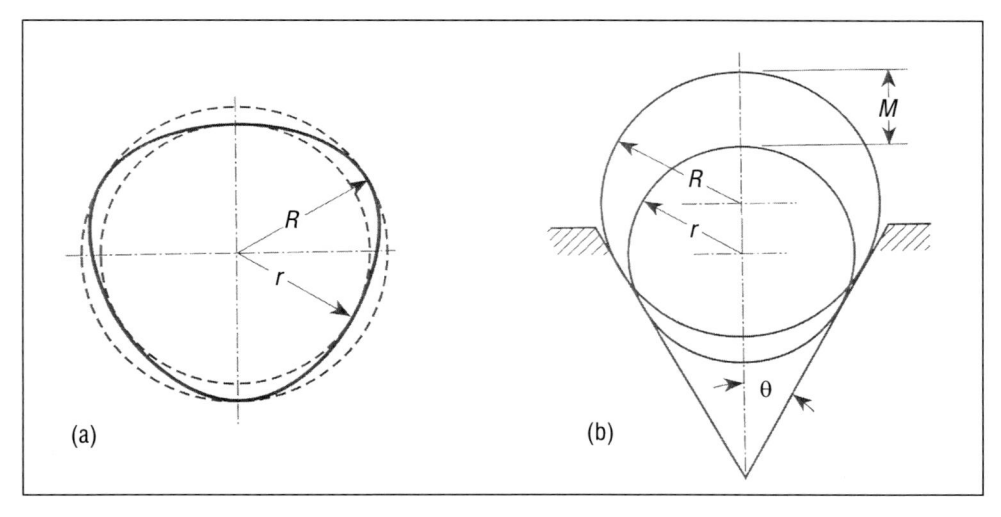

Fig. 2

158   Quality Gaging Tips

1) Included angle:
2ø = 180 - 360/n
Where "ø" equals half the included V-block angle, and "n" equals the number of lobes.
2) Multiplication Factor:
M = (R + R csc ø) - (r + r csc ø)
or
M = (R - r) (1 + csc ø)
Where "M" equals the measured difference (as shown on the indicator), "R" equals the radius of the circumscribed circle, "r" equals the radius of the inscribed circle, "ø" equals $1/2$ the included angle of the V-block, and "R - r" equals the specified radial variation (tolerance) for the part.

This can be simplified. The table below, worked out from these formulae, gives the required multiplication factor and included V-block angle for 3-, 5-, and 7-lobed conditions, which represent the majority of the odd-lobed conditions found in normal machining practice.

Condition	Multiplication Factor	Included V-Block Angle
Three lobes	3.00	60°
Five lobes	2.24	108°
Seven lobes	2.11	128°34'

Let's use a specific example. The part is specified to be "round within 0.0001 in. variation in radius." We know the part was produced on a CNC lathe using a 3-jaw chuck, and we have previously determined (possibly through the use of a circular geometry gage on a short-term basis) that the process generates a 3-lobed condition. Therefore, we use a gage having a V-block fixture set up with a 60-degree included angle. To determine allowable variation in radial out-of-roundness, multiply 0.0001 in. × 3. Any TIR larger than 0.0003 in. is therefore out of tolerance.

Bearing in mind that such measures are only approximate, these techniques provide a good, practical means to determine out-of-roundness on the shop floor.

■

Circular Geometry Gaging Means More Than Roundness

Previously, we looked at the measurement of out-of-roundness. But roundness is far from the only circular geometry specification that machinists may be required to meet and, therefore, to inspect. Let's look at some of the other parameters. As we describe them, refer to the figure to see how each is indicated on part print callouts.

Roundness involves no datum: it is evaluated relative to the part profile itself, using one of the four methods discussed in the previous column (Maximum Inscribed Circle, Minimum Circumscribed Circle, Least Squares Center, or Minimum Radial Separation). Eccentricity, in contrast, is measured relative to a datum, which is the center of part rotation, as established by the spindle of the geometry gage (or by a part feature defined as the datum that has been centered on the spindle). Eccentricity is the distance between the center of the reference circle used to calculate out-of-roundness, and the datum. As the part rotates 180° around the datum axis, the center of the reference circle is displaced by twice the eccentricity value: hence, <u>concentricity</u> is twice eccentricity. Both eccentricity and concentricity may be measured for features lying in a single plane, or in two planes.

<u>Circular runout</u>, another datum-referenced measurement, measures the radial separation of two concentric circles whose common center is the datum, and which entirely enclose the part profile. Circular runout is the result of the combination of two form-error factors: out-of-roundness, and out-of-concentricity. The two factors may be additive or may cancel each other out, depending on vector directions. <u>Circular flatness</u> (of a flange, for example) may be specified at an indicated radius, and measured in a circular trace. This is a datum-free measurement that uses either a minimum-zone or least squares calculation, similar to those used in roundness measurements.

Circular flatness can be used as the basis for <u>plane parallelism</u> measurements. Care must be taken, however, in reading and interpreting callouts correctly. The statement "A is parallel to B" (within a specified tolerance) implies that surface B is the datum. Any out-of-flatness present in this surface is ignored, while out-of-flatness in surface A is included in the calculation. The gage user cannot treat the two surfaces interchangeably. If one excludes out-of-flatness of both surfaces, the measurement is defined as <u>parallelism plane runout</u>.

Symbol	Name
○	Out-of-Roundness
E, ⊙⊙	Eccentricity
◎	Concentricity
↗	Runout (Circular)
▱	Flatness (Circular)
∥	Plane Parallelism, Linear Parallelism
∥↗	Plane Parallelism Runout
⊥	Perpendicularity, Squareness
⊥↗	Plane Perpendicularity Runout
⌭	Cylindricity
—	Straightness
①, ⌯	Coaxiality
↗↗	Total Runout

In order to measure a number of squareness-related parameters, a vertical datum axis must first be established by measuring the roundness of the part at two planes, thus creating a part axis between the centers of the two reference circles. After normalizing the part axis to the gage spindle's axis of rotation, the horizontal surface in question is gaged at a specified radius, and normalized to the datum axis. <u>Perpendicularity</u> includes the out-of-flatness of the horizontal surface, while <u>perpendicularity plane runout</u> ignores out-of-flatness. <u>Squareness</u> is defined as half the plane runout value—in other words, it measures only from the center of the part's rotation to the indicated radius, while perpendicularity plane runout measures the deviation across the entire circle.

All of the parameters above can be measured on so-called "roundness" gages, which do not provide a means for precision vertical movement of the gage head. "Cylindricity" gages, on the other hand, incorporate precision reference surfaces in the gage head positioning axes, permitting measurements of a number of additional parameters.

Cylindricity is a useful parameter that provides an overall assessment of part roundness, taper, and straightness. Because it is not possible to measure every point on a three-dimensional surface, part profiles are taken at a number of planes, then combined into a single cylindricity value. Statistical analysis, and experience, may be required to establish the number of sample profiles needed for an accurate measurement.

Cylindricity gages can also be used to measure the straightness of an ID or OD surface on a vertically oriented workpiece, by keeping the part stationary and traversing the gage head up or down. Straightness can then be used as the basis for linear parallelism measurements, comparing opposed ID or OD surfaces, or comparing an ID surface to an OD surface.

We haven't space here to describe additional, complex parameters such as coaxiality and total runout. The main point, however, is that numerous parameters have been developed in order to control the functionality of parts across a wide range of possible configurations and applications. Make no assumptions when gaging part geometry: be sure you understand what the parameter means before you try to measure it. A couple of useful reference sources are: Geo–Metrics II by Lowell W. Foster (Addison Wesley Publishing); and the ANSI B89.3.1 standard for out-of-roundness measurement.

■ ■ ■

Geometry Gaging: Part 1

Circular geometry gaging is becoming increasingly common in machine shops of all sizes. Once confined to metrology labs in large companies, the technology has been migrating into the hands of machinists since the early 1990s.

My colleague, Alex Tabenkin, says this is due to several factors, one of which is the improved ease of use of geometry gages. The use of personal computers as gage controllers has greatly simplified the complex calculations required for some measurements, while improved user interface devices, such as touch screens and joysticks, have made the gages easier to set up and program. Other factors aiding the migration to the shop floor include substantial price reductions and enhanced ruggedness.

But perhaps the most important influence is coming from design engineers and quality control departments, who are demanding control over geometry as a means to improve the functionality of parts. To oversimplify the issue: parts that are supposed to be round work better if they are truly round, within certain tolerances. But most metrology labs can't keep up with the demands for inspection on a production basis, so it has been necessary to turn the function over to QC inspectors or machinists, who are increasingly using the gages to check and control their own work.

But let's back up and take a close look at the equipment. As shown, the basic elements of the gage are a turntable, a gage head stand, a gage head, and a data acquisition device in the form of a PC. We'll go through each in order, considering some of the engineering and purchasing considerations involved.

The heart of the turntable is a precision spindle: in the gage pictured, the spindle is

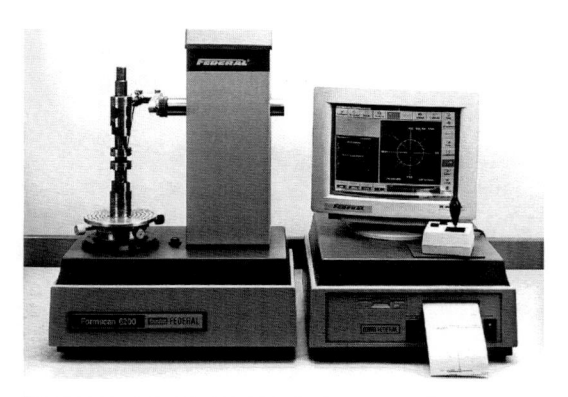

The basic components of a modern geometry gage are (left to right): a tilt-centering table mounted on a precision spindle; a gage head stand supporting an electronic, lever-type gage head; and a data acquisition device, here in the form of a PC.

supported by an air bearing, which promotes accuracy and reduces friction to essentially zero. The bearing requires a compressed air supply, but this is justified by the spindle's axial and radial accuracy of ±1 microinch.

Mounted on the spindle is a tilt and centering table. The user manipulates the table's knobs to adjust the orientation of the part, so that it is concentric and coaxial with the spindle. A three-jaw chuck or a custom holding fixture may be mounted on the table as a further aid to setup, although neither is shown here.

The gage head stand positions the gage head both horizontally and vertically. In the present instance, the stand is motorized in both vertical and horizontal planes, with positioning controlled by the joystick device on the computer. If speed of setup is not an important consideration, some money can be saved by ordering a gage equipped with a manually operated stand.

In the gage shown, the stand is not a precision device: it is simply a means to position the gage head. Gages with non-precision stands are capable of measuring several geometry parameters: roundness (also called out-of-roundness or circularity), concentricity, circular runout, circular flatness, perpendicularity, plane runout, top and bottom face runout, circular parallelism, and coaxiality. For convenience's sake, such gages are referred to as roundness gages.

Other gage head stands incorporate precision vertical and horizontal slides, with known levels of straightness and linear positioning accuracy. The slides thus serve as reference surfaces, allowing the gage to perform additional measurements where gage head positioning is critical. These are: cylindricity, total runout, vertical straightness, and vertical parallelism. Such instruments are called cylindricity gages. As you might expect, the additional precision engineering adds to the cost of the system, so potential users must carefully analyze their requirements before choosing between "roundness" and "cylindricity" gages.

The electronic lever-type gage head is just barely visible in the photo, near the top of the workpiece with its stylus pointing almost directly at the viewer. The stylus on this gage head can be easily replaced: a useful feature, as new ISO standards specify the use of different stylus tip radii as a function of the diameter of the feature being measured. Some users may also need an extra long stylus, to measure inside a deep bore, for example.

The data acquisition device takes the signal from the gage head and converts it into numerical results, or displays it in the form of a polar or linear chart. With the exception of the joystick, all controls are located on the touch screen, which allows the user to initiate setup programs, select the desired geometry parameter, choose various analytical and statistical functions, and store and print out results.

Geometry Gaging Part II:
Four Methods Of Measuring Out-Of-Roundness

We have introduced the subject of circular geometry gaging by looking at the instrumentation, and we noted that one reason for the recent proliferation of geometry gages is the use of personal computers as gage controllers. The PC has greatly simplified geometry measurements by speeding up the calculations involved. Now, let's proceed to the most common geometry measurement, and the basis for most circular geometry parameters: roundness, also known as out-of-roundness or circularity. As we'll see, even "simple" roundness has benefited greatly from the processing power of the modern PC.

Ideal roundness, according to ANSI standard B89.3.1, is "the representation of a planar profile all points of which are equidistant from a center in the plane." Out-of-roundness, then, is "the radial deviation of the actual profile from ideal roundness," and the out-of-roundness value (OOR) is "the difference between the largest radius and the smallest radius of a measured profile; these radii are to be measured from a common point".

To measure out of roundness, then, it is necessary to compare the part profile to an ideal circle or datum. But since the part profile itself isn't round, how do you locate the ideal circle?

Four methods are in common use. Many modern geometry gages offer users a choice. Typically, the user selects the required method, and then initiates the measurement on the gage. The gage rotates the part and collects data, which it presents in the form of a polar chart. Then the computer controller uses one of the following methods to locate the center of the reference circle:

Maximum Inscribed Circle (MIC): the center of the largest circle that can fit within the measured polar profile. This method is used only for geometry measurements of inside diameter features.

Minimum Circumscribed Circle (MCC): the center of the smallest circle that fits around the measured profile. This method is used only for outside diameter features.

Least Squares Center (LSC): the center of a circle, of which the sum of the squares of the radial ordinates of the measured profile is the least possible number. This method is used for both ID and OD features.

Minimum Radial Separation (MRS): the center of two concentric circles, which, with the least possible separation, contain all points of the profile. This method is also used for both ID and OD features.

Different part applications typically call for different measurement methods. For example, when the geometry of an inside diameter is specified, the presence of burrs, dirt, and other "high points" on the ID are typically of critical concern, while low points (e.g., scratches) are not quite as important. Accordingly, inside diameters can be measured using the MIC method, because it is quite sensitive to high points, and relatively insensitive to low points. In other words, a burr will cause a significant shift in the location of the center, while a scratch will cause only a minor shift.

On the other hand, scratches tend to be of greater functional concern on outside diameter parts, while burrs tend to be of less importance. The MCC method, which is sensitive to scratches, and insensitive to burrs and dirt, therefore has advantages for measuring outside diameters.

The MRS method is quite sensitive in equal measure to both positive and negative asperities (i.e., burrs and scratches) and typically generates the largest OOR value of the four methods. The LSC method, in contrast, is relatively insensitive to extreme asperities of both kinds, and therefore generates the most stable center and the smallest OOR values of the four methods. As both of these methods react equally to positive and negative asperities, they tend to be useful for measuring mating ID and OD parts. And because most ID parts do have a mating OD part (and vice versa), the MRS and LSC methods are in more frequent use than the MIC and MCC methods.

OOR values may differ by as much as 10-15 percent from the same measurement data, depending on the method used. Inspectors must refer to the part print callout before firing up the gage.

The use of the proper reference circle has importance beyond just OOR measurements: many other parameters are based on roundness and the location of the circle's center, and they too will be influenced by the method selected. Concentricity, circular runout, total runout, coaxiality, and cylindricity are all affected. Now, aren't you glad the gage controller will run the calculations for you? (Some gages even allow the user to store the data, and then apply the different measurement methods on a post-process basis.)

If the part print callout doesn't specify the method, MRS is the default, according to ANSI, even though LSC is in more common use. My colleague Alex has qualms, therefore, about the use of a default. If the method isn't shown in the callout, you never know if the engineer intended that the default method be used, or if he simply forgot to take it into consideration. Alex therefore recommends that engineers use the ISO convention, which requires that the method be specified. It's certainly not a lot of extra trouble to add the information to the callout, and it may help avoid unnecessary confusion.

■ ■ ■

Air Rings, CMMS And Supermikes

A major aerospace customer complained that the air-ring gage we sold him was inaccurate. How did he know, I asked. Because, he said, he checked the measurements against a coordinate measuring machine and a supermicrometer he had in the shop. The CMM and the supermike agreed with one another closely, while measurements on the air gage differed from them by as much as .0004", ergo...

"Send me a few samples," I told him, "and we'll check them in our lab, where results are good to one millionth. Then we'll know exactly what size they really are, and which gage is at fault."

The lab identified at least part of the problem even before they put them on a gage. "Where are the witness marks?" they wanted to know. "Where, exactly, were these parts measured?"

"What difference does it make?" asked the customer. "They're simple OD cylinders."

In fact, it makes a lot of difference. If a part is slightly out-of-round, then the measuring method you choose will influence your measurement. A CMM, for example, will tend to average out errors of geometry and waveform. A supermike might give you the Min,

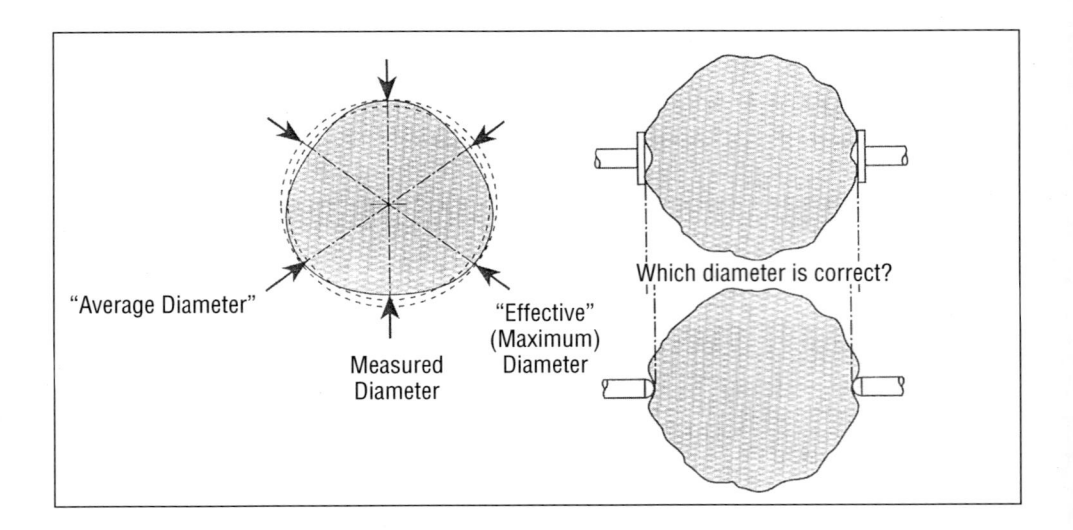

"Average Diameter"
Measured Diameter
"Effective" (Maximum) Diameter
Which diameter is correct?

the Max, or somewhere in-between, depending upon precisely where the measurement is made. The performance of an air ring can also vary between Min/Max and average reading, depending upon the number of jets, the part's geometry and surface finish, and the position of the part in the gage. None of them are necessarily wrong.

In this particular case, all three gages were giving accurate readings, but each one was measuring different dimensions. In the lab, we found that the parts exhibited geometry errors of as much as .0003", in addition to a small amount of waveform error. By measuring at different locations on the parts, the manufacturer sometimes picked up on that variation, and sometimes missed it. Simply by measuring from a consistent datum, we brought the air gage readings to within 50 millionths of the other two gages.

Instead of asking "is it accurate?" we should be asking "is it appropriate?" Most gages are accurate as delivered from the manufacturer, but every gage embraces certain limitations and assumptions. When selecting a gage or a gaging method, it is essential to establish a clear objective: Do you want to account for, or ignore, variation due to geometry, waveform, and surface finish? Do want to know the maximum OD of a part, or the minimum OD, or the average OD?

The answer to these questions depends upon the application. As a hypothetical example, consider a spool valve assembly, in which the bore is a perfect cylinder, and the spool itself has a slight three-lobed condition. The overall (average) diameter of the spool may determine the efficiency of the valve, but its maximum diameter will determine whether the two parts can be assembled or not. It's up to the user to determine which is the critical measurement, and then select the measuring tool most appropriate to the task.

Many gages offer a certain degree of flexibility. For example, it may be possible to specify the arrangement of jets in an air ring to automatically give the Min/Max, or average reading. Likewise, it may be possible to program a CMM to account for geometry factors. But before you can do either, you have to know what you want to measure.

Not surprisingly, this situation is paralleled by the factor of surface finish. Air gages tend to average, or ignore, surface roughness—up to a point. A supermike, measuring on the "peaks," will tend to maximize its effect, while a CMM will randomize peaks and valleys,

Section J: Geometry

generally giving an average. In the aerospace manufacturer's case, we found that surface finish accounted for the remaining difference in readings between the gages.

And if that isn't complicated enough, here are two more factors you might want to consider: 1) The geometry of the gage's sensitive contact and holding fixture may affect measurements. 2) Masters are also machined parts that are subject to the influences of geometry and surface finish.

Why didn't we worry about this stuff before? Because even as recently as 10 years ago, tolerances were generally looser. But as tolerances get tighter, variations in part geometry and surface finish exert proportionally more influence on our measurements.

Section K

Air

You Won't Err With Air

It is perfectly natural that machinists should have an affinity for mechanical gages. To a machinist, the working of a mechanical gage is both straightforward and pleasing. Air gages, on the other hand, rely on the action of a fluid material, the dynamics of which are hard to (shall we say?) grasp. But air gaging has many advantages over mechanical gages and should be seriously considered as an option for many applications.

Air gages are capable of measuring to tighter tolerances than mechanical gages. The decision break point generally falls around 0.0005 inch; if your tolerances are tighter than that, air gaging provides the higher resolution you will need. At their very best, mechanical gages are capable of measuring down to 50 millionths, but that requires extreme care. Air gages handle 50 millionths with ease, and some will measure to a resolution of 5 millionths.

But let's say your tolerances are around 0.0001 inch and mechanical gaging would suffice. Air still provides several advantages.

The high-pressure jet of air automatically cleans the surface of the workpiece of most coolants, chips, and grit, aiding in accuracy and saving the operator the trouble of cleaning the part. The air jet also provides self-cleaning action for the gage plug itself. However, the mechanical plug-type gages can become clogged with cutting oil or coolant and may require occasional disassembly for cleaning.

The contacts and the internal workings of mechanical plug gages are subject to wear. There's nothing to wear on an air plug except the plug itself, and that has such a large surface area that wear occurs very, very slowly. Air gages consequently require less frequent mastering and, in abrasive applications, less frequent repair or replacement.

On some highly polished or lapped workpieces, mechanical gage contacts can leave visible marks. Air gaging, as a non-contact operation, won't mark fine surfaces. For the same reason, air gaging can be more appropriate for use on workpieces that are extremely thin-walled, made of soft materials, or otherwise delicate. Continuous processes, as in the production of any kind of sheet stock, rolled or extruded shapes, also benefit from non-contact gaging.

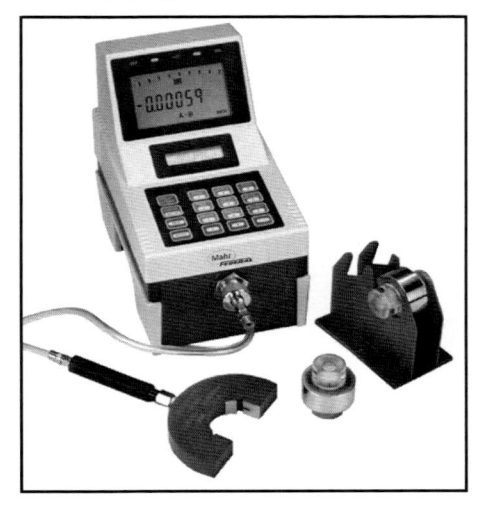

Air equipment can save time in almost any gaging task that is not entirely straightforward. Air plugs with separate circuits can take several measurements simultaneously on a single workpiece, for example, to measure diameters at the top and bottom of a bore for absolute dimensions, or to check for taper. Jets can be placed very close together for measurements of closely spaced features.

Air plugs are available (or can be readily engineered as "specials") to measure a wide range of shapes that would be difficult with mechanical tools. Examples include: spherical surfaces, interrupted bores, tapered bores, and slots with rectangular or other profile shapes.

It would be possible to design a fixture

gage with a number of dial indicators to measure several dimensions in a single setup, such as diameters of all the bearing journals on a crankshaft. But a fixture gage using air gaging will almost inevitably be simpler in design and fabrication, easier to use, less expensive and more accurate.

Because of the relative simplicity of fixture design, air gaging is especially suited to relational, as opposed to dimensional, measurements, such as squareness, taper, twist, parallelism, and concentricity.

Air gaging isn't perfect, though. Its high level of resolution makes air gaging impractical for use on workpieces with surface finish rougher than 50 microinches R_a because the readings would average the highs and lows of the rough surface. Most important, air gaging has relatively high initial cost, so it is usually reserved for large production runs. Clean, compressed air is also expensive to generate and must be figured into the equation.

In general, however, air gaging is the fast, economic choice for measuring large production runs and/or tight tolerances.

Flexibility Of Air Gaging

In a previous article, I touched on several applications where air gaging is particularly practical. These include relational, as opposed to dimensional, measurements, such as distance between centers, taper and concentricity. Along with high resolution and magnification, speed and repeatability, air gaging exhibits great flexibility.

Air gages are often simpler and cheaper to engineer than mechanical gages. They don't require linkages to transfer mechanical motion, so the "contacts" (jets) can be spaced very closely, and at virtually any angle. This allows air to handle tasks that would be difficult or extremely expensive with mechanical gaging.

Gaging the straightness and/or taper of a bore is a basic application that benefits from close jet spacing (see Fig. 1). All it takes is a single tool with jets at opposite sides of the gage's diaphragm. The gage registers only the difference in pressure between the two sets of jets, directly indicating the amount of taper.

The concept can be applied to a fixture gage, to measure several diameters and tapers in a single operation. This fixture would be much simpler than a comparable gage equipped with mechanical indicators, each one outfitted, perhaps, with a motion transfer linkage and retracting mechanism. And the air gage could have an electrical interface to single in- or out-of-tolerance conditions with lights—much quicker to read than dial indicators. From one basic concept, we have just described at least four options (there are more). How is that for flexibility?

Note the basic principle: air circuits operating on one side of a diaphragm measure dimensions; circuits on opposing sides measure relational measurements or differences between features. The beauty of the concept is that you

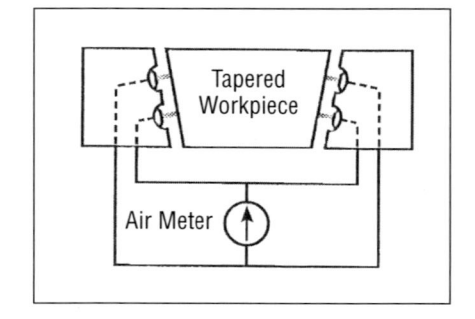

Fig. 1

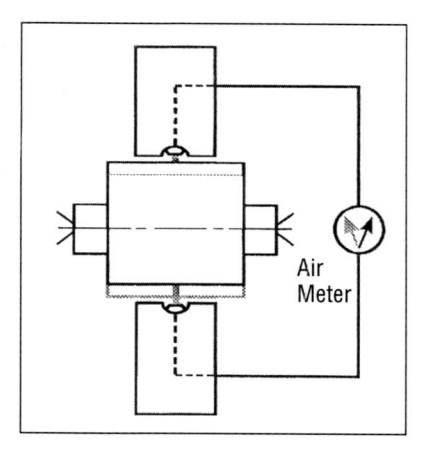

Fig. 2

can choose to ignore dimensions while seeing only relational measurements, and vice versa. You never have to add, subtract, or otherwise manipulate gage data. It can all be done with direct reading.

Consider, for example, the fixture gate, which allows the shaft to be rotated. Two jets on opposing sides of the journal, acting on the same side of the diaphragm, will accurately measure the diameter of the journal, even if it is eccentric to the shaft. As the shaft is rotated, the air pressure increases at one jet, but decreases at the other one. Total pressure against that side of the diaphragm remains constant, so we obtain a diameter reading.

If the two circuits operate on opposite sides of the diaphragm (see Fig. 2), the gage reflects not the total pressure of the circuits, but the difference between them. Higher pressure on one side or the other, therefore, indicates the journal's displacement from the shaft centerline.

Naturally, one can position the jets and circuits to measure both features at once, and add another set to check for taper. Could you design a mechanical gate to accomplish all this at once? Perhaps, but at the cost of mechanical complexity.

The same principles apply to gages designed to measure the squareness of a bore (see page 114, "Checking Bores For Ovality And Taper"), the included angle of a tapered hole, or the parallelism (bend and twist) and distance between centers of two bores (such as a connecting rod's crank and pin bores). Bore gages with the proper arrangement of jets can turn checking barrel shape, bellmouth, ovality, taper and curvature into quick, one-step operations.

Contact-type air probes, which use air to measure the movement of a precision spindle, provide the major benefits of air gaging (high resolution and magnification) where the use of open air jets is impracticable for use in measuring.

Open jets can't measure against a point or a very narrow edge. The narrowest jet orifices are 0.025 inch, and these require a somewhat broader workpiece surface to generate the necessary air "curtain" to read accurately. Contact-type air probes, however, can gage with point or edge contact. Rough surfaces (above 50 microinches R_a) will baffle an open jet, but present no problems to a contact probe. And where open jets are limited in range to 0.003 inch to 0.006 inch measuring range, contact probes can go out to 0.030 inch (at some loss in sensitivity, however). Contact probes are often mounted in surface plates to measure flatness. Depth measurement of blind holes is another common use for long-range probes.

You needn't be a gage engineer to appreciate the flexibility of air gaging, or to understand how it can simplify gaging tasks. Just remember that virtually all kinds of measurements—both dimensional and relational—can be performed with air, and that the more complex the task, the more air recommends itself.

Air Is Free, But Not Carefree

Air gaging represents the method of choice for most high-resolution measurements on large production runs. While quite durable and reliable compared to mechanical gages, air gaging is not carefree.

Accurate air gaging requires proper maintenance of the tooling, and vigilance over the air supply. Although the factory air supply may not be under the gage user's control—compressors and air lines may be shared by dozens of other users—the gage user must ensure that the air reaching his gage is clean, dry, and fairly stable in pressure. Tooling, on the other hand, is directly under the gage user's control, and he is responsible for its maintenance.

Proper tool maintenance simply means keeping it clean and dry inside and out. Contaminants such as chips, dirt, coolant, and cutting fluid may be picked up from workpieces, while water and oil are likely to come from the air source itself. Although the airflow tends to clear out most air passages on its own, some contamination may occur in the master jet or measuring jets. Even accumulations of only a few millionths of an inch can throw off a measurement. The gage must be inspected and cleaned when necessary. Repeated mastering that produces varying readings is a good indication of dirty jets.

Shop air is difficult to keep clean and dry. Air dryers are not entirely adequate. The very act of compressing air produces moisture, and a compressor's need for lubrication inevitably generates some oil mist in the line. Oil and water mist can actually act as an abrasive

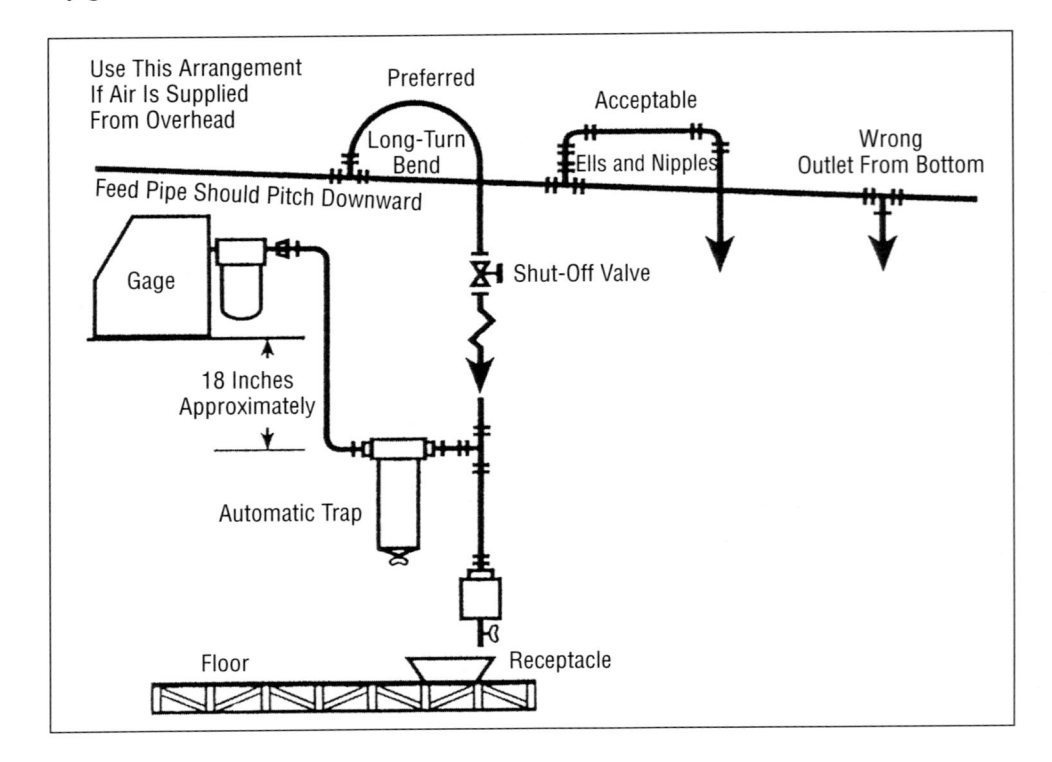

Section K: Air

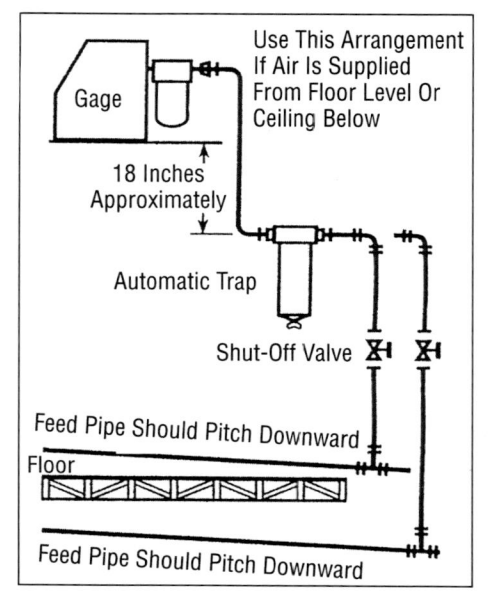

and cause part wear over long periods (for example, the Grand Canyon), so don't leave the gage on overnight. Our near-term goal, however, is simply to prevent mist from entering the gage and fouling the jets. To do this, we must employ proper air line design to intercept it before it enters the meter.

Air main lines should be pitched down from the source, with a proper trap installed on the end. Feed lines should also be equipped with traps. Take air from the top rather than the bottom of the mains, so that moisture doesn't drain into the feed. Long, gentle bends on feeds are preferable to hard angles and close ells. Bleed air lines before connecting gages to them. Gages must always have a filter in place when operating, and this should be changed when it becomes saturated.

So, enough about moisture and that oil. Let's talk about air.

Air leaks are another common cause of air gage inaccuracy. To test, cover the measuring jets tightly with your fingers and observe the indicator needle. If it's not stationary, check all fittings, tubes and connectors for leaks. Simple.

Most factory air lines run at about 100 psi, but depending upon the demands of other air users, this can fluctuate widely. Properly designed air gages operate reliably over a wide range—some as wide as 40-150 psi—so a certain amount of fluctuation is acceptable. Other gages are more sensitive and must be isolated from fluctuations by using a dedicated or semi-dedicated air line. To check the sensitivity, simply leave a master in place on the tool and observe the indicator for movement as other air line users perform their normal tasks.

If large pneumatic equipment is being used on the same air line, surges over 400 psi might be generated that could blow out the built-in regulator and damage the gage itself. Again, isolation of the line is the solution.

Tight, clean and dry: the requirements of air gaging aren't very different from mechanical gaging after all.

■ ■ ■

Air Gaging For Itty-Bitty Holes

Conventional air gaging for measuring inside diameters is typically limited to a minimum size of about 0.060"/1.52 mm: below that, it becomes difficult to machine air passages in the plug tooling, and to accommodate the precision orifices or jets. But air gaging is among the most flexible of inspection methods, and with a simple change of approach, it can be used to measure very small through holes, below 0.040"/1 mm in diameter.

Most air gages measure back pressure that builds up in the system when the tooling is placed in close proximity to a workpiece. In the case of bore gaging, a smaller bore means closer proximity of the part surface to the jets: this results in higher air pressure, which the gage comparator converts into a dimensional value.

A few air gages measure the rate of flow through the system rather than back pressure: as tool-to-workpiece proximity decreases, flow also decreases. The flow principle can be effectively applied to measure very small through holes, even on air gages that were designed to operate on the back pressure principle. Rather than installing tooling at the end of the air line, the workpiece itself is connected to the line. Smaller bores restrict the flow of air more than larger ones. Thus, the workpiece essentially becomes its own gage tooling. This approach works on all common types of back pressure gages: single-leg gages requiring dual setting masters, as well as differential-type gages, which typically use just one master.

Airflow is proportional to the bore's cross-sectional area, but area varies with the square of diameter. Gage response in this setup, therefore, is non-linear. Nevertheless, this rarely causes problems, because the range of variation to be inspected is usually very small, and the gage is typically set to both upper and lower limits using dual masters or qualified parts. If numerical results are required, specially calibrated dials may be installed on analog comparators, while some digital comparators allow software correction.

Back pressure air gages operating as flow gages for small holes have been used in a number of specialty applications, ranging from fuel injection components to hypodermic needles. Often, all that is required is a special holder that allows the part to be attached quickly and easily, with a good air seal. Air pressure and flow stabilize quickly, making this method efficient for high-volume inspection.

Like other forms of air gaging, flow-type measuring of small through holes is extremely adaptable. It has been used to measure IDs as small as 12 microinches/0.3 micrometers, and as large as 0.050"/1.27 mm. Range of measurement can be as long as 0.006"/0.15 mm, and discrimination as fine as 5 microinches/0.125 micrometers. In some cases, where the hole is so small that airflow is negligible, bleeds may be engineered into the system to boost total flow to a measurable level. On the other hand, excessive flow through large bores may be brought down into a measurable range by engineering restrictors into the system.

Some parts, including some fuel injection components, have two holes sharing a common air passage, and require that the holes be measured twice: once simultaneously, and once independently. To accommodate this requirement, special two-station air gages have been designed, where the first station connects the air flow through both holes, while the second station only connects the air circuit to one of the holes, and blocks the other.

Many other methods exist to inspect small holes. Some applications are served well by microscopes and optical comparators, although neither is well suited to high-volume production applications, and both are limited in the part configurations they can accept. Go/no-go gaging with precision wires is also practical only for very low volume tasks. The air gaging method described here often requires a modest level of application engineering, and occasionally a custom dial face or gaging fixture, but it lends itself well to high-throughput inspection of very tight tolerances. Some users have experienced up to 300 percent increases in efficiency compared to other methods.

In discussing air gaging in past columns, we've often emphasized its flexibility. With it, one can measure a wide range of dimensional characteristics, including inside and outside diameters, feature location, thickness, height, and clearance/interference. Air can also be used to measure geometry characteristics such as roundness, squareness, flatness, parallelism, twist, and concentricity. And we've seen how air gages can measure very deep bores, blind holes, and counterbores. The use of air gaging to inspect very small through holes is yet another example of the tremendous adaptability of this relatively simple, but very cost-effective technology.

Choosing The Right Air Gage

Air gaging has many advantages as an inspection method. It is quick and easy to use, requiring little skill on the part of the operator. It is highly adaptable to measuring special features for both dimensional and geometric tolerances, ranging from simple IDs and ODs to taper, flatness, and runout. With different tooling readily installed on the gage display unit, it can be highly economical. And as a non-contact form of measurement (in the sense that there are no hard contacts), air gaging is useful for measuring delicate or flexible surfaces, and for monitoring the stability of continuous processes such as drawing and extruding.

Once the decision has been made to use air, the user can choose between three basic types of gages, each operating on a different principle. These are: the flow system; the differential pressure or balanced system; and the back pressure system.

In older flow-type gages, air flows upward through a graduated glass column containing a float. Exiting the column, it flows through a tube to the tooling, where it exits through precision orifices or jets. Flow increases with clearance between the jets and the workpiece. When clearance is large, air flows freely through the column and the float rises. When clearance is small, airflow decreases and the float descends. Flow systems are not very popular in production environments, because they do not readily provide high magnification, and tend to be sensitive to clogging.

The other types of air gages measure pressure, not flow. As clearance between the jets and the workpiece increases, pressure decreases. In the back-pressure system, both the pressure meter itself and the bleed (i.e., zeroing circuit) are "Tee'd" off of a single air line—at the end of which is the tooling. Back pressure systems are often called "dual master" systems: with a relatively short range of linearity, two masters are required to set the upper and lower tolerance limits.

In the differential system, the line is split into two legs. The bleed is at the end of one leg; the tooling is at the end of the other; and the bellows-type meter is located between the two legs. When pressure in both legs is equal, the meter is centered at zero. When a change in distance between the tooling and the workpiece causes pressure in the measuring leg to increase or decrease, the bellows reacts accordingly, and this is reflected on the meter. Differential systems offer linear response over a relatively long range: a single master is therefore sufficient to establish the zero point and still assure excellent accuracy on both the plus and minus sides.

Both differential and back pressure systems are very well suited to production gaging applications, for different reasons. Differential systems are capable of higher magnification and discrimination; are easier to use because of greater tool-to-part clearance and the requirement for only one master; and are more stable. Back pressure systems offer lower cost, adjustable magnification, and greater interchangeability of tooling between manufacturers. See the table for a summary of benefits associated with these gages.

Back Pressure Vs. Differential Air Gaging

Back Pressure (Dual Master) Gages
- Adjustable magnification; tooling flexibility.
- Less costly tooling.
- Higher air pressure: cleans part surface more effectively.
- Two masters provide greater traceability.
- More manufacturers; wide compatibility.

Differential (Single Master) Gages
- Higher magnification, discrimination; longer range.
- Greater tool-to-part clearance reduces wear, speeds usage.
- Better stability, dependability; no drift. Better for automatic control applications, and data collection for SPC.
- Single master makes gage easier, quicker to set.

■ ■ ■

Checking For Centralization And Balance Errors

Air gaging is often referred to as a non-contact form of measurement. This is accurate, to the extent that there's no metal-to-metal contact between a *sensitive* gage component and the workpiece. Nevertheless, air gage tooling—including air plugs for inside diameter measurements—does generally come in contact with the workpiece, and may show wear after several thousand measurements or years of use. (The comments here are equally applicable to electronic plug gages.)

When, due to wear, the clearance between the gage and the workpiece exceeds the design clearance, centralization error results. The air jets then measure a chord rather than the true diameter of the part. As the distance between the chord and the bore centerline increases, we begin to see measurement inaccuracy. Another form of error occurs when the jet centerline is not on the plug centerline. In this case, the plug will always measure a chord of the part.

How much centralization error is allowable depends upon both the diameter of the workpiece and the dimensional tolerance specification. Obviously, looser tolerances can "tolerate" more measurement error. But equal amounts of misalignment will cause greater centralization error in a small bore than in a large one. (For details on calculating the misalignment tolerance based on allowable centralization error, refer to page 44, "Central Intelligence.")

For a gage to function properly, all the jets in an air plug (or ring) must have a common recess depth and orifice diameter. But recesses and orifices may become clogged with con-

Fig. 1 Fig. 2 Fig. 3
A   A   A
B   B   B

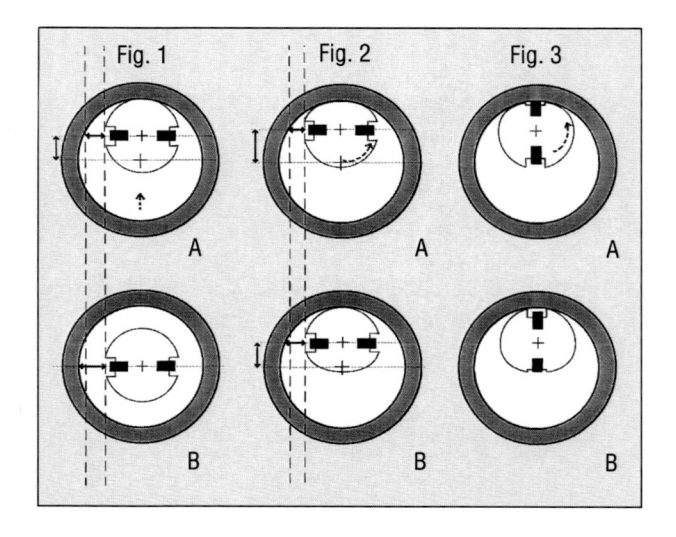

taminants, damaged through accident, or worn unevenly through very heavy usage. This causes another form of measurement error, called "balance error."

It is easy to inspect for these various forms of error. Certain types of gage usage will demonstrate characteristic wear patterns. In hand-held gaging applications, wear usually occurs fairly evenly around the circumference of the plug. If, however, the plug is horizontally mounted on the front of an air gage display, then the top surface will probably experience the most wear.

If wear is expected to be fairly regular around the circumference, start by securing the gage horizontally, with the jets also horizontal. Place a master on the plug, release it, and note the reading (Fig. 1A). Then carefully raise the master until it contacts the lower surface of the plug. If the plug is worn, the readout will change as the measurement moves from a chord, through the maximum diameter (Fig. 1B), to another chord. Wear may be considered excessive if the reading changes by an amount equaling 10% or more of the part tolerance.

If the gage is stationary, and the jets are normally oriented horizontally, wear might be expected on the uppermost surface of the plug between the jets. Again, place a master on the plug, release it, and note the reading (Fig. 2A). Then rotate the plug 180° and note the reading again (Fig. 2B). If the surface in question is worn, the second reading will be higher.

If the jets are normally oriented vertically, then the top jet, or its recess, are of the greatest concern. The test is performed as above, by placing the master, noting the reading, rotating 180°, and reading again (Figs. 3A, 3B). Alternately, the test can be performed by lifting the master ring, as in the first procedure described. Normally, two-jet air plugs automatically balance themselves when one of the jets is closer to the workpiece than the other—as is the case here, where the master is allowed to rest on the upper jet. But if one jet or orifice is damaged or worn, this test will demonstrate the gage's inability to maintain that balance.

Plug gages tend to be highly durable, because they contact the workpiece across a broad surface area. But that doesn't mean you can ignore the possibility of poor centralization or balance. Include these tests in your annual gage calibration program.

The 3 D's Of Straightness Plugs

We have touched on the many different applications of air gaging in a number of articles: size, match gaging, and form applications such as taper. In this column, we'll spend some time with air straightness plugs and how they work.

The typical out-of-straightness condition is seen as a "bow" form within the bore, one that was introduced as part of the manufacturing process. The air straightness plug attempts to measure the depth of its curve. Usually this out-of-straightness condition is all within one orientation along the axis of the hole.

Design

A typical air plug has four measuring jets in two opposing sets—two near the middle and two near the ends, as seen in Figure 1. This design allows the plug to look at both extremes of the bow condition.

There are no rules for the exact positioning of the jets relative to each other—as is sometimes the case with taper or squareness checks. Nor are there any ratios involved. The air jets at the extreme of the plug are positioned to inspect for the out-of-straightness condition, which is usually specified over the total length of the bore. But, in order to understand how a straightness plug works, we have to take a quick look at the various combinations of jets typical in air tooling.

Differential Measurement

A two-jet plug is a differential measuring system. Imagine a two-jet air plug inside a master ring with the indicator reading zero. Now move the plug so that one jet is pressed against the side of the ring. This increases the back pressure on one jet, and decreases it on the other. But the indicator reading does not change, because the combined pressure of the two jets remains the same. However, when you insert the plug into a smaller or larger test part, the pressure does change, and the gage reads the differential.

An extension of the two-jet air plug is a four-jet system. Here, four jets are added together and if the plug is moved in any direction, again an average (differential) reading is made. The four jets see four pressure changes and add them all together. If there is a change in any of the measured dimensions, the total—and the reading on the indicator dial—changes.

The four jets are normally at the same level or plane on the plug. In theory, the four jets could be moved any-

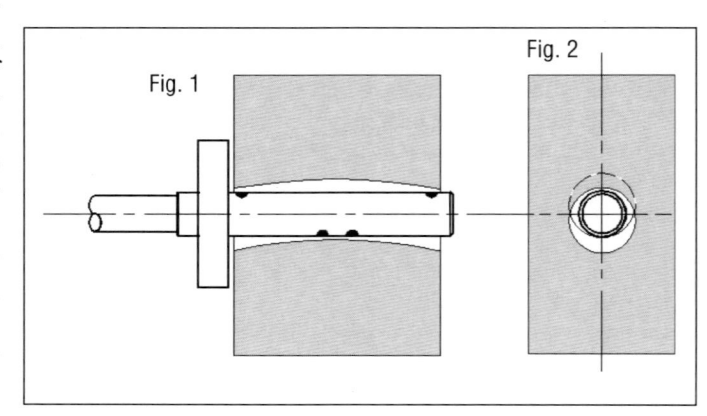

Fig. 1

Fig. 2

Section K: Air 177

where along the length of the plug independently, and if they are positioned at 90 degrees to each other, they will measure an average diameter of the bore. But if we move the jets so that two are on the same side at the extremes of the plug, and the other two are moved to the center on the other side of the plug, we have the straightness plug described above. If the part being measured is perfectly straight, and the plug is moved up and down, it acts like a two-jet piece of tooling. The two jets on the top are offset by the two jets on the bottom, and the result is no change on the display. But if the bore is not perfectly straight, the combined pressure changes, and the differential is shown on the instrument.

Dynamic Measurement: But don't go thinking you've got it down so fast. If the straightness plug is simply inserted into the bore, the display shows a number. What that number means is the real question. Straightness needs to be made as a dynamic measurement, very similar to a squareness check. With both of these form errors, the out-of-form condition is at its maximum along the axis at two positions, 180 degrees opposite each other.

Again, looking at Fig. 1, we can understand what the inner and outer jets are looking at. When the jets are in line with the bow (jets up and down) the jets are seeing either their Max or Min reading depending on the orientation. When moved 180 degrees, the inner and outer jets reverse roles and the same value is seen, and the plug is working in its differential mode.

But as the plug is *rotated* through 180 degrees exploring the bore, the pairs of jets will experience a maximum clearance and then find the minimum clearance, usually at right angles to each other. This difference between the maximum and minimum value is the out-of-straightness condition as seen over the total length of the plug measurement length.

You can also think of it this way. If you looked at the bore from the end, and drew a line around the extremes of its path, you would end up with an ellipse, Fig. 2. If you had an ellipse in a hole, a two-jet air plug would be capable of measuring the variation in size by rotating it in the part. Think of the straightness plug as a stretched-out air plug with four jets doing the same thing. With this in mind you have a pretty good understanding of what the straightness plug is doing.

The air straightness plug, though it is a little more complicated then a standard air plug, still maintains the same advantages of a standard air plug—easy to set up, easy to use, and high precision results.

Sliced Bread And The Limits Of Air Gaging

"Air gaging is the greatest thing since sliced bread," a friend once told me. And he was right—air gaging is good. It's fast, high resolution, non-contact, self-cleaning and easy to use. For use in a high-volume shop, it's hard to beat. But that begs the question, "If air gaging is so good, why would you ever consider going back to contact type gaging?"

The answer is that while air gaging does provide all of the benefits listed above, it and everything else that obeys the laws of physics, has some limitations. There are, in short, some trade-offs and for every advantage you gain in the measuring process with air, you will have to pay the price and sacrifice something else. The real question is, "What are those limitations and how can you best work with them?"

Air gaging gives you a fast measurement device that provides superior reliability in the dirtiest shop environment. But you give up things like measurement range and a clear delineation of surface. Air gaging has about 10 - 20 percent of the range of a typical electronic transducer with similar resolution.

The response of air to surface finish, however, is more complicated. Think of an air jet. The measurement 'point' is really the average area of the surface the jet is covering. Now consider the finish, or roughness, of that surface. The measurement point of the air jet is actually the average of the peaks and valleys the jet is exposed to. This is not the same measured point you would have if a contact type probe is used. This difference is a source of real gaging error, and one, which is most often apparent when two different inspection processes are used.

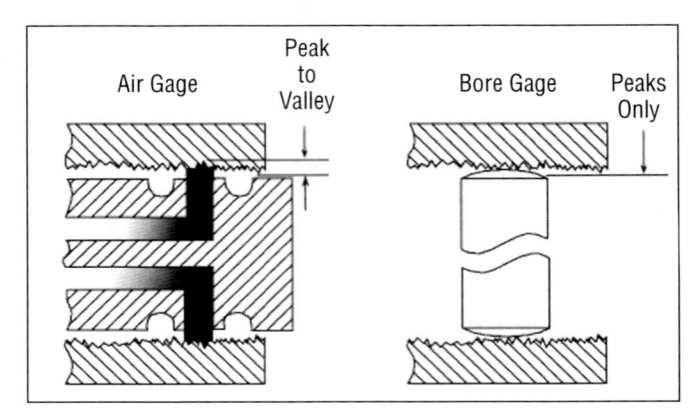

Unlike the contact probe on a bore gage, the measurement point of an air jet is actually the average of the peaks and valleys to which the jet is exposed.

For example, let's say we have a surface finish of $100\mu"$ on a part, and we're measuring with an air gage comparator and two-jet air plug that has a range typically used to measure a 0.003" tolerance. The typical gaging rule says you should have no sources of error greater then 10 percent of the tolerance. In this example, that's 0.0003". If we used this plug on the $100\mu"$ surface, the average measuring line is really $50\mu"$ below the peak line. Double this error with two jets and you get 0.0001" or 30 percent of the allowable error. That's pretty significant and air would probably not be a good choice for this part. As a general rule, the limit for surface finish with an air gage is about $60\mu"$, but it really depends on the part tolerance.

This source of error should also be considered when setting the plug and comparator to pneumatic zero. If the master and the part have similar surface finishes, then there is little problem. Most master rings are lapped to better then a $5\mu"$ finish. However, if the gage is now used on a $200\mu"$ finish part, there would be significant error introduced. For most applications, there should be no more then $50\mu"$ difference between the master and the part the gage is measuring. Even this can be significant if the tolerance of the part is as little as 0.001".

In some applications air gaging can be the best thing since sliced bread. In others, you can get in trouble with the butter. When measuring porous surfaces, narrow lands, and areas extremely close to the edge of a hole, stick with a fixed size, mechanical plug with probe contacts.

Advances In Air Gaging

The advantages of air gaging are numerous. It is fast and accurate for production environments. The gage is easy to use, and helps clean the part by blowing dirt away.

Over the years, the tooling for air gaging has remained basically the same: steel tubes or rings with precision orifices that set up a pressure/distance curve when in use. As the orifice is restricted, flow is reduced and pressure builds up in the system. This principle can be used to monitor the distance between the air jet and the part surface. Orifices can be arranged in complex configurations to indicate form or relationship errors to the user. Typical applications include average diameter, taper, squareness, center-distance, and straightness. These fixed tooling components provide fast results for the user.

Unlike the tooling, however, air-gaging displays have changed dramatically over the past 50 years. Initially, the air meter consisted of a water-filled tube. Changes in air pressure from closing the tooling orifice caused the water in the tube to rise or lower, and part of the gage setup process involved filling the tubes. Ironically, one of the biggest sources of error in an air gage today is the very thing that used to be needed to make the old gages work—water in the air lines.

The next generation of air gage displays used a float in a tapered tube to monitor the flow of air to the orifice. Some air gages of this style can still be obtained. It's a simple design that works, but it's difficult to keep clean and resolution may be lacking for some applications.

Mechanical amplifiers were next employed to monitor the expansion and contraction of a bellows assembly as the pressure in the air system increased or decreased. Differential air gage circuits came into play and provided an opportunity to manufacture tooling and air displays to a high level of performance. These systems held such tight standards that a single master would place the air system into the "sweet" spot on the air pressure/distance curve.

Electronic air amplifiers took amplification one-step further. The first air-to-electronic transducers were LVDT's used to monitor the position of the same type of bellows used in mechanical gages. Other systems use silicon-based transducers to monitor the pressure change. Both methods provide high resolution and fast response. Combined with the electronics in a gaging amplifier, they can make the gage a lot smarter and provide dynamic functions that can average readings and combine air signals for even more complex measurements.

One area that has not seen much in the way of advances until recently is the portability of the air gage. Hand tools are used all the time in the quality/process control function to bring the measurement to the part. With air gaging, the need for the piping and mechanical or electronic amplifiers always meant that the displays were rather big. For small parts this was no problem: the parts could easily be brought to the gage. For larger parts, the tooling could be attached to a hose and brought to the part, but there was a potential problem in that the operator had to watch the tooling as he placed the plug in the part, then turn around to search for the air display in order to interpret the value.

In recent years, electronics have allowed even air displays to become smaller. There are now air gages on the market that make the display portable, as part of the air tooling itself. With this configuration, the air display is right in front of the operator, ready to be read, and there is no searching for the old air meter. Like the amplifier version, these portable air gages combine electronic transducers with today's digital indicators, to provide selectable resolutions, tolerances and data collection capabilities.

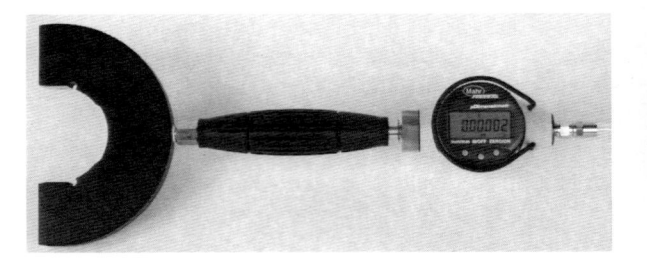

Air gaging was thought to have reached its peak many years ago. But in reality, there is an increasing demand for its application. As tolerances have gotten tighter on the shop floor, air gaging is often the only way to make the check fast and easy for the operator. With portable air gaging, that check can now be made right at the part.

Section L

Automatic

Process Control Gaging

Most often, this column addresses gaging as an element of the QA inspection process that occurs after a part is machined. This may be performed on the basis of SPC sampling, or 100 percent automated gaging. However, other approaches to dimensional gaging can serve other functions. Machine tool evaluation, for example, involves gaging the machine, to determine its potential for producing accurate parts. And in addition to mere inspection, post-process gaging can be used to maintain control over the process, either by manually compensating for observed drift, or through automatic feedback loops. Gaging can also occur in process, to control the machine in "real time."

Process control gages actively participate in machine control, whether on a post-process, in process, or even a pre-process basis. Process control gaging is typically used when variables exist that may affect the stability of the process and cause frequent out-of-tolerance conditions. Such variables can include: tool wear; growth or deformation of the machine tool or workpiece from internal "environmental" sources such as heat or vibration; and external environmental influences such as vibration from nearby machines. Process control gaging can help achieve tight tolerances in spite of such variables, and thus eliminate the need for intensive operator intervention, help improve productivity, and reduce scrap.

Where process drift tends to be gradual, post-process control gaging is often a practical approach. Under this regime, parts are fed directly from the machine tool into the gage. When the gage measures deviation away from the nominal specification and toward the upper or lower tolerance limit, it generates a feedback message, signaling the machine tool's NC controller to apply compensation, so that each successive part is brought closer to nominal.

Post-process control gaging may not provide sufficiently rapid response to some process variables. With grinding in particular, the tool may wear substantially during the production of a single part. Or a process may be inherently unstable, and thus unpredictable. In such cases, in-process gaging may be the best approach.

During in-process gaging, part size is monitored constantly to provide continuous, real-time control over the machine. Gaging is typically performed by electronic gage heads, which may be placed in direct contact with the workpiece, or may act against a moving part of the machine's drive mechanism that has a direct relation to part size—for example, the ram on a centerless grinder. The gage and machine controller can be programmed so that the machine initially runs at a rapid feed rate, then slows down to creep feed when most of the grind stock has been removed, and stops when the nominal dimension is reached.

In some instances, process variables resist even in-process control—for example if the part is subject to significant, unpredictable thermal expansion. Pre-process control gaging may be the answer in such cases. The part is gaged prior to machining, and the gage instructs the machine on how much material to remove before grinding begins.

The electronic gage heads used in process control applications are rugged, well shielded against contamination, and designed to work with coolant and chips flying all around. Even so, gage heads do become worn and damaged. Most process control gages therefore rely upon standard gaging components, which are readily and economically replaced.

Contact-type gage heads may leave slight wear marks on some parts, particularly those made from aluminum or other soft materials. If such marks are unacceptable, either on cosmetic or functional grounds, air gaging can be used for in-process control. Air leaves no marks, and it offers the same potential for microinch accuracy as electronic gage heads.

The other main element in the control gage is the amplifier/controller. (Air gaging interfaces with electronic amplifiers through air-to-electronic converter modules.) Modern systems offer higher resolution and accuracy, and faster response than earlier versions, permitting very rapid machining rates on high precision parts. Some new systems also offer convenience features such as automatic zeroing and offset (tool wear compensation) capabilities, switchable inch/metric digital displays, and a choice of comparative or absolute size measurements.

Process control gaging has advanced spectacularly since the days of "snap-on" mechanical grinding gages, when the machinist would adjust feeds and speeds in "real time" according to the movements of the needle on a dial indicator. Even compared to the amplifier systems of just a few years ago, the new grinding gages are faster, more accurate, and easier to use. If your system predates the '90s, it may be time to look into an upgrade. And if you're not using process control gaging at all, check it out: it may be just the answer to a troublesome production problem.

■ ■ ■

Automated Gaging

Every shop should have this gaging "problem." A Tier 2 automotive supplier was required by his OEM customer to perform 100 percent inspection on parts for dimensional tolerances and several other characteristics. The parts were aluminum forgings for air conditioning compressor pistons, and the inspection requirements included dimensional tolerances for bend and twist, and checks for fill at four positions, presence of two radii, and flash removal at two positions. The problem was the size of the contract: the supplier was obligated to deliver, and thus to inspect, nearly 100,000 parts per day.

Some gages lend themselves to faster throughput than others. Generally, the more specific the gage's purpose, the more quickly it can be operated. For example, fixed-size bore gages (i.e., plug gages) are quicker than adjustable (rocking) bore gages. If the task is more complex—for example, inspecting multiple diameters on a part such as a crankshaft—then a purpose-built fixture gage will allow faster throughput than a selection of snap gages, or the use of surface plate methods. But there comes a point when even the most narrowly focused, manually operated gage must give way to automation. It's either that, or hire a whole roomful of inspectors to keep up with production.

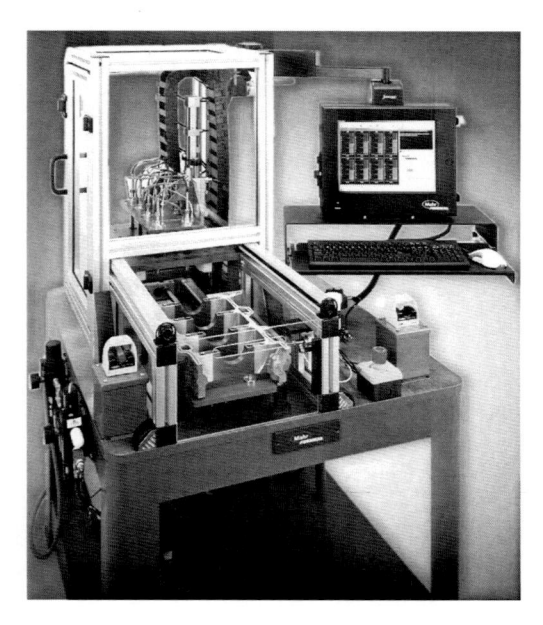

Automated gaging is usually cost-justified in applications where a part must be inspected every 45 seconds or less. This figure includes not just gage operation, but the entire gaging cycle. While the part measurement itself may take only three seconds, the complete cycle includes at minimum: placing the part in the gage; operating and reading the gage; and removing the part from the gage. Other required actions may include: recording the measurement; sorting parts into appropriate categories by size; and removing rejects from the lot.

Many additional variables influence the speed at which a part can be measured, and hence influence any gaging setup, whether manually operated or automatic. These include: the number of features to inspect; the need for a dimensional reading versus simply go/no-go results; how measurement data will be used (e.g., for export to SPC, or for direct process feedback); the level of accuracy required; and whether gaging occurs in-process or post-process.

With so many variables in play, it is hardly surprising that automatic gages can rarely, if ever, be bought "off the shelf." In the case of the Tier 2 supplier, we custom-engineered a fully automatic gage, capable of inspecting one part every 3.5 seconds (i.e., 1,030 parts per hour, or 24,720 parts per day). Four identical units were built and installed, providing total throughput of 98,880 parts per day.

Reliable parts handling was obviously one of the most important engineering considerations. Accordingly, the gage was designed to make use of the most reliable parts-handling mechanism ever developed: gravity. Parts feed in at the top of the machine, sliding down a 45° chute to a dead stop, where a proximity switch senses the part and triggers a locking mechanism. An air-driven cylinder then raises the holding fixture, and a nest of electronic gage heads descends until it contacts the part. The gaging device traverses the part, checking for true position, material fill, flash, and the presence of radii. Bend and twist are checked as independent features by comparing position and diameter measurements at opposite ends of the piston.

When the inspection is complete, the holding fixture descends and the part is released to drop down the exit chute. Out-of-spec readings trigger an escapement, which diverts bad parts into a reject bin, while good parts pass straight through to the next production process.

Gage head signal conditioning, data processing, and data storage are controlled by a gaging computer, from which measurements are downloaded daily. Programmable logic controllers (PLCs) control all of the gage's logic functions. The systems have proven to be extremely reliable, operating around the clock for months between downtime for preventive maintenance.

To equal the throughput of the automated gage, the Tier 2 supplier would have to keep 13 human operators working around the clock, at a minimum throughput of one part every 45 seconds per person. Even at minimum wage, and assuming no coffee breaks or sick days, it wouldn't take long before those human operators, each with a manually operated gage and a master, started to look pretty expensive.

Few machine shops face throughput requirements even close to this, but any shop involved in a large production run can usually benefit from some form of specialized gaging, to make inspection easier and faster. And for larger shops where throughput requirements are very high, and the production run will last for a year or more, customized, automated gaging may be the only practical approach to inspection.

Machine Compensation

Ever since electronics first made their way onto machine tools, machine builders and users have been seeking ways to integrate gaging results, to achieve some level of "automatic" process control. Certain causes of dimensional variation in machined parts—tool wear, for instance—occur gradually. Measuring parts for variation provides a means to efficiently adjust the machine's position settings, to "compensate" for tool wear and other changes.

In the first use of electronics for compensation, parts were gaged by hand, and then the operator would press a button on a stepper control attached to the machine tool. One button moved the machine a fixed amount in one direction, while another moved it the same amount in the opposite direction.

The next generation, which integrated electronic gages, took the concept further with automatic feedback control. When the gage sensed that a part had reached or exceeded an approach tolerance, it would send a signal to the machine's controller to compensate. Again, the compensation was a fixed amount each time, and this was known as incremental compensation.

As microprocessors and computers were incorporated into both gaging and machine tools, the simplicity of incremental compensation was replaced by the sophistication of absolute compensation, in which the machine's position is adjusted by the exact amount that is optimum for the process. If desired, compensation can be triggered when part dimensions are drifting just a little bit off nominal, rather than waiting for them to approach tolerance limits.

Computers or microprocessors run algorithms that determine the present level of the process, and look for trends, steps, or other statistical features. While there are a number of different statistical schemes, all use size data from a number of consecutively machined workpieces to establish the current level of the process. A popular, basic scheme is to simply take the average size of the most recent parts. Averaging a large number of parts tends to minimize the influence of normal part-to-part variation, and reduces the number of compensations performed. Small sample lots, on the other hand, allow the machine to respond faster to process changes. Other algorithms, which may be based on any known method of statistical process control, can be far more sophisticated.

There are two basic hardware options for modern, automated machine tool compensation. Microprocessor-based CNC/gage interfaces accept input from a variety of electronic gages and gaging amplifiers, and communicate with the CNC via RS232. These are usually panel-mount devices that are pre-programmed to perform a wide range of standard gaging/control actions, with integral keypads that allow users to set approach and tolerance limits, select algorithms, define actions, and so on. The second option is a gaging computer, which offers higher-level capabilities to store or modify algorithms, analyze and utilize data simultaneously from a larger number of inputs, program more complex actions, and store and communicate data. They can be readily interfaced with most modern CNCs and other production equipment.

For example, a large agricultural equipment manufacturer uses a gaging computer to maintain control over gear blank production in fully automated workcells. A robot loads a part into an NC lathe, then removes the half-turned part, turns it around, and loads it into another lathe, which turns the other half. The robot then places the part on an automated gage which measures several ID and OD dimensions. After gaging, the robot stacks the parts on pallets.

All gage functions are controlled by the gaging computer, which allows the manufacturer to switch instantly between eight different part numbers. The computer compensates the lathes based on the average deviation from the three most recent parts, and shuts down the cell instantly should any dimension fall out of tolerance. The computer also accepts data from temperature sensors in the gage, and performs additional machine compensation for thermal influences.

The use of gages for automatic machine compensation can improve overall quality and productivity, reduce scrap, and minimize manpower requirements. It should be seriously considered for all long-running automated or semi-automated applications where dimensions must be maintained within close tolerances.

Semi-Automatics—The In-Between Gages

The faucet has been opened a little and you've just received a long-awaited contract to produce 10,000 large trunnion caps for a manufacturer of earth moving equipment. Despite the joy, you realize you have a problem. The machines will be in place and ready to run the part shortly, but you haven't given much thought to the gaging.

Manual gaging is not going to cut it because it is too slow. Operator-introduced variation may be too large for the tolerance, and there are too many parts to inspect over a relatively short period. On the other hand, the size of the job just won't justify the expense of a fully automatic gage.

There is something in-between that can solve the problem: a semi-automatic gage—one that makes multiple checks, classifies the part, and can mark or stamp it for identification. These are frequently, manually loaded/unloaded, but can also incorporate a disposal system.

In certain applications, these features offer the user a number of distinct advantages. Since part of the gaging cycle is automatic, semi-automatic gages are much faster than completely manual, individual gages. And, when manually loaded, they can handle workpieces that may be difficult and costly to feed or orient manually. Manual loading of the part also permits visual inspection for scratches, discoloration and unclean finishes prior to the gaging process.

Semi-automatic gages will check a relatively large volume of parts quickly and accurately, enabling the inspector to keep up with production by taking over the gaging function. Many manually loaded gages can be operated at speeds close to one part per second. Disposal is automatic, eliminating operator interpretation or sorting errors. Most semi-automatic gages are controlled by a small gaging computer that takes over the complete gaging function—positioning gage heads, moving the part, collecting data, marking the part, and disposing of it in the correct class. Operator fatigue and misclassification can be a big problem when handling many parts. Since the semi-automatic gage is tireless and consistent in its decision-making, many of the operator-influenced problems go away.

Today there are many choices in semi-automatic gages, and design time is not as long as is sometimes perceived. A semi-automatic, built with off-the-shelf gaging components and supplemented with motion control and gaging computers, can be designed quickly and delivered ready to meet its gaging challenge.

Now if we could only get it to put the parts in boxes and deliver them to ……

Section M
Hand Tools

OD Gaging Can Be A Snap

Insert a workpiece into a snap gage, and you'll understand how these extremely effective, fairly simple tools for checking precision ODs got their name. You have to push deliberately to get the part past the leading edges of the anvils. But once you've overcome the 4 $^1/_2$ lbs. of spring force, the part slips suddenly back against the backstop, contacting it with a good, healthy "snap."

Snap gages can be hand-held to measure workpiece ODs still on the machine, or can be mounted on stands for use with small parts. The heart of the tool is a simple C-frame casting, and measurements rely upon a direct, in-line, 1:1 transfer of motion. These factors make snap gages simple, reliable, and fairly inexpensive.

The earliest snap gages, of the fixed, or Go/No-Go variety, did the job well enough so that thousands are still in use. But fixed gages have some distinct liabilities. For one thing, you need a different gage for every dimension you wish to check.

But the biggest shortcoming of the Go/No-Go snap gage is that it tells you nothing about your process. You know if you're within tolerance, but you can't see if you're gradually getting larger or smaller, and so adjust the process accordingly—until you're out of tolerance.

Then some clever engineer (not I) replaced the upper anvil of a fixed snap gage with a dial indicator, and the whole scenery changed. *Indicating* snap gages are able to measure to the limits of resolution of the indicator, and as they are comparative gages (i.e., they read to zero), they give the machinist a window on his OD machining process.

But the high cost associated with a different gage for every dimension was not solved until the introduction of the *adjustable* indicating snap gage. With a typical range of adjustment of one inch, the adjustable snap can eliminate dozens of tools from the shop. Adjustable snaps are still comparative gages: the adjustable jaw is set to a master or a stack of gage blocks, and the indicator is "zeroed" before gaging begins.

With a standard dial indicator installed, the measuring range of an adjustable snap gage is typically .020", with a resolution of .0001". But there's no rule that says an adjustable indicating gage has to have a mechanical dial indicator. One can specify digital indicators, air probes, or electronic probes—they all use the same standard $^3/_8$" diameter mounting. With an electronic probe and amplifier, you can achieve resolution of 10 microinches, for tolerance measurements tighter than .0005".

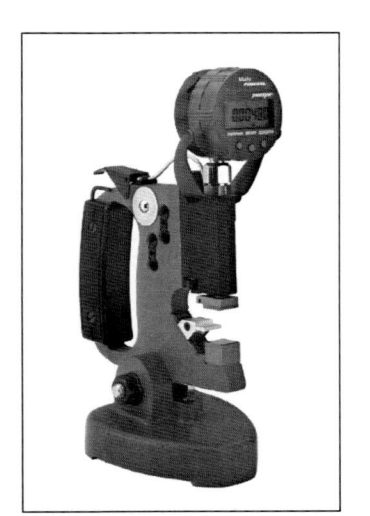

Usually no modifications are necessary to retrofit an indicating snap gage with the probe or indicator of your choice. It's a simple, in-house job, and it's very possible to take an entire shop's worth of old mechanical indicating gages and refit them with digital indicators or electronic probes. And these can be tied into SPC or other computer-based quality systems.

Adjustable indicating snap gages can be modified to accommodate special applications. Extra-large C-frames can be built to measure ODs up to 48". Anvils can be side-relieved, chamfered, or straddle-milled to provide access in difficult part profiles. Blade-type contacts can be used to measure diameters or grooves right up against perpendicular surfaces.

Usage is particularly simple and straightforward, but, as with any gage, there are a number of principles of operation and maintenance one must observe to obtain accurate measurements and long life.

■ ■ ■

Using Adjustable Snap Gages

Let's continue the previous discussion about adjustable, indicating snap gages. The principles of care and usage for these simple OD measuring tools are straightforward. Because the body of the gage—the C-frame—is a rigid piece of metal, most of the "care and feeding" tips are concerned with the gage's anvils. That's where most of the precision lies.

Make sure the gage is suited to the application. The anvils should be narrower than the part being measured to avoid uneven wear on the measuring surfaces. If you repeatedly gage narrow parts on a broad anvil, you can wear grooves that may not be picked up by mastering. You can get away with a small number of too-narrow parts, but if you're doing a production run, buy different anvils or modify the existing ones.

Anvils can be straddle-milled or side-relieved to fit into grooves or recesses, or to ensure they're narrower than the workpiece. The edges can also be chamfered. This is important when measuring a diameter immediately adjacent to a perpendicular feature—for example, a crank throw on a shaft. There's usually a fillet where two surfaces come together, and if you put crisp, sharp-edged anvils right up against the perpendicular, you'll measure the fillet instead of the critical dimension. Another way to phrase it is: Don't check diameters next to perpendicular surfaces—unless you've got the right anvils.

Regularly check the anvils for wear. Look for scratches, gouges, unevenness, pitting, rust, etc. If problems are detected, the anvils can be removed and their surfaces ground

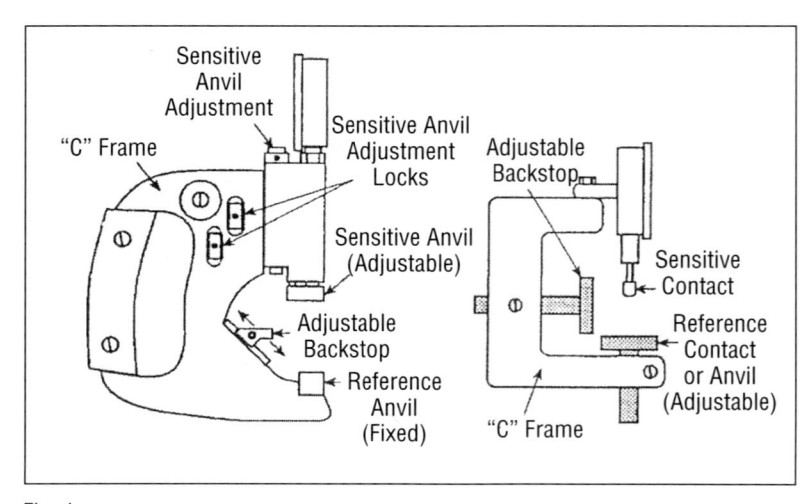

Fig. 1

Straddled Milled   **Side Relieved**   **Chamfered Edges**

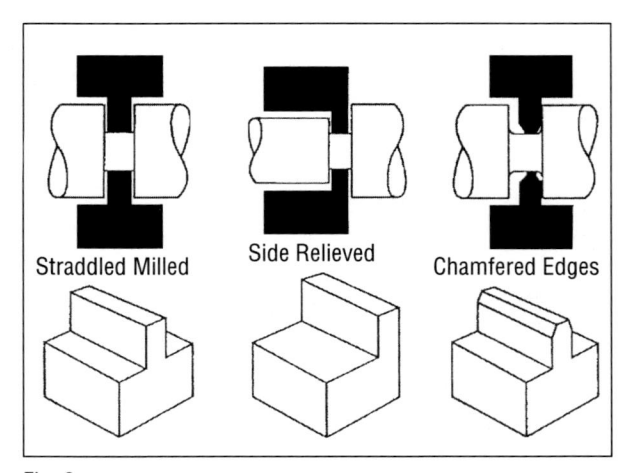

Fig. 2

and lapped. Check periodically that the anvils are parallel. This is essential if you've removed the anvils for maintenance or replacement. To check for parallelism, place a precision wire or a steel ball in sequence at the front, back, left and right edges of the anvils. Compare the indicator reading for each of the edges.

If you detect an out-of-parallel condition and you haven't just replaced the anvils, you've probably dropped the gage. While it's recommended to have the manufacturer tweak it back into shape, many shops can handle this in-house. Remove the fixed, lower anvil, and carefully file the seat in the indicated direction. Go slowly—just a few gentle licks—then re-mount the anvil securely and test again for parallelism. Leave the seating of the upper, moving anvil alone.

Observe the basics of good gaging practice: check regularly for looseness of components, keep the gage clean, protect against rapid changes in temperature, and master regularly. For large production runs, it makes sense to purchase a master disc the size of your part. For small runs, use stacked gage blocks. Make sure you've wrung them properly and observed the other basics of block care and usage.

Adjustments on indicating snap gages are few and simple. Set the backstop so the diameter of the workpiece is roughly centered on the anvils: it's not a critical adjustment. To adjust the gage's capacity, turn the knurled nut that moves the upper anvil/indicator assembly up or down. Move the upper anvil until the indicator zeroes itself against the master. Then, before you tighten the locking nut(s), turn the adjusting nut very slightly in the opposite direction to release the torque on the lead screw. This may seem insignificant, but any amount of tension will relax itself over time. Then lock it down, master the gage, and check for repeatability several times before you start measuring.

Wide anvils normally ensure that the gage seats itself squarely on the part. But if you're using narrow blade-type anvils to check narrow grooves, you have to hold the gage as steady as you can, squaring it up by eye. Offset blade anvils also impose side loading, which can further reduce repeatability. To accommodate these shortcomings, lower-resolution dial indicators are usually used with blade anvils: .005" resolution is typical, compared to .0001" on most snap gages.

For large gages that weigh several pounds, the spring pressure on the upper anvil may be insufficient to achieve repeatability in a hand-held situation. There's a simple solution to this one: turn the gage upside down and allow the weight of the gage to rest on the fixed anvil instead. Then just rotate the bezel on the dial indicator, so it reads right side up.

Inspecting Multiple Diameters

This column has discussed circular geometry gages at some length, and pointed out the rapidly growing popularity of these extremely useful instruments. It is important to bear in mind, however, that most inspection gaging of round features is still performed with traditional indicator gages.

When measuring multiple diameters on shafts and similar parts with two or more cylindrical features, fixture gages incorporating multiple indicators offer convenience, speed, and economy. By building multiple gaging stations into a single fixture, it is possible to eliminate the expense of duplicate work-holding devices. The gage user need only fixture the part once, and can quickly scan across the indicators or readouts, thus saving time and effort over the use of multiple gages that each measure just one feature.

Gaging a single outside diameter is among the simplest of inspection tasks, but gaging multiple ODs simultaneously in a fixture gage can be deceptively tricky. Even when using a gage designed specifically for the part in question, it is possible to get diameters, roundness, and concentricity mixed up. Different gages or setups are required to properly measure each of these parameters.

In gaging fixtures, workpieces may be held in V-blocks or between centers, or may be stood on end. Some workpiece feature or features must rest securely against reference points on the fixture, to establish the proper position relationship between the gage's sensitive contact and the workpiece. In the case of V-blocks as references, end-journals on the workpiece are commonly used.

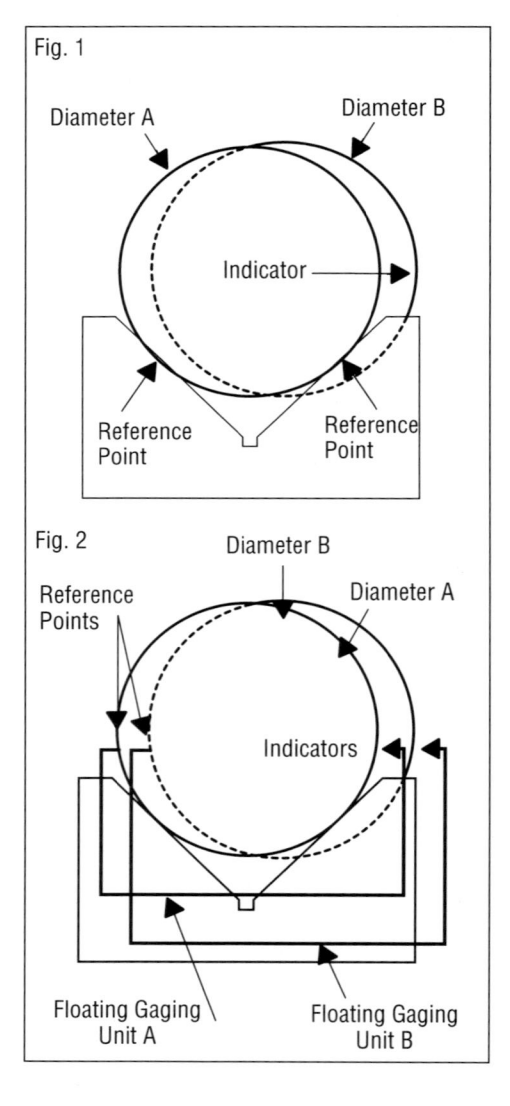

Problems arise when cylindrical features are assumed to be concentric with the end journals but are not in fact. Diameter A establishes the relationship between the part and the gage. Because Diameter B is not concentric with Diameter A, its position relative to the gage contact is unknown, and it changes as the part is rotated. Diameter B therefore cannot be measured for diameter or roundness. On the other hand, this setup can be used to assess the concentricity of Diameter B relative to Diameter A. Furthermore, because Diameter A is properly referenced, we could measure it for size and roundness simply by building a second indicator into the gage. (For simplicity, we've shown only two features on the part, but some gages may measure many more.)

If one wishes to measure Diameter B for size or roundness, the gaging unit must incorporate a pantograph-style gage or similar mechanism that is free to "float" around the non-concentric feature, while maintaining a fixed relationship between the reference and the indicator. (Air gaging with opposing jets on a single circuit may also be used, if out-of-concentricity will fall within a limited range. As the eccentric feature moves closer to one jet, it moves away from the opposite one, so that total pressure in the system remains constant and the gage reading remains unchanged.)

We stated above that measuring a part with a single diameter is among the simplest of tasks. There are conditions, however, where single-diameter work that is significantly out-of-round will appear perfectly round when using a micrometer, snap gage, or any other two-point method (such as that shown in Fig. 2). This is especially so with centerless-ground parts, in which three or five lobes may appear evenly spaced around the part's diameter, so that high and low points are all diametrically opposed. In this situation, the diameter remains nearly constant, even though the radius may vary significantly.

A circular geometry gage is required to detect and understand this condition. Thereafter, it is possible to perform approximate out-of-roundness inspection on a production basis using indicator gaging. The part is staged on a V-block, and a simple indicator is positioned over the V-block's centerline using a height stand or similar comparator device. As the part is rotated in the V-block, the Total Indicator Reading (TIR) is noted, then multiplied by a constant. For three-lobed parts, use a V-block with a 60° included angle, and multiply TIR by 3.00; for five lobes, use a 108° angle, and multiply TIR by 2.24; for seven lobes, use a 128°34' angle and use a 2.11 multiplication factor. This method provides sufficient accuracy for roundness measurements within a few thousandths. For greater accuracy, a true circular geometry gage may be required.

Calipers: Ideal For Measurement On The Go

Although it has been around for a long time, the caliper is still an extremely versatile and useful tool for making a wide range of distance measurements (both ODs and IDs). While micrometers are more accurate, they have a limited measurement range (typically several inches). The caliper, on the other hand, can span from two inches to four feet, depending on the length of the scale. External measurements are made by closing the jaws over the piece to be measured, while internal measurements are made by opening up the inside diameter contacts.

Three Types

There are three different types of caliper, which may be found today in a machinist's tool chest.

Vernier. The vernier caliper was the original design and is still the most rugged. Graduated much like a micrometer, it requires the alignment of an etched scale on the vernier plate with an equally spaced scale running the length of the tool's handle. Skillful alignment of the tool and interpretation of the reading is necessary to achieve the measurement tool's stated accuracy.

Dial. A dial caliper is the second-generation caliper. Similar to the construction of the vernier caliper, this style replaces the vernier scale with a dial indicator. The indicator is fixed to the moveable jaw, and engaged with toothed rack on the body of the unit. The dial, which is typically balanced (i.e., can move in either plus or minus directions from zero), may be graduated in either inch or metric units.

The dial caliper is a dual-purpose tool for making either direct or comparative measurements. To make a comparison, first measure the reference dimension and set the dial indicator to zero. Then measure the compared dimension. The indicator will show how much the compared dimension varies from the original (plus or minus).

Another useful feature of the dial caliper are jaws, which slide past each other to allow contact points or depth rod extensions to fit into narrow openings for small ID measurements.

Digital. In the last 20 years the digital caliper has made its way onto the shop floor. The latest designs provide many numerous electronic features, which make the device easier to use, but add little in the way of cost. These include: easy switching between inch and metric units on the readout, tolerance indications, digital output to electronic data collection systems, zero setting anywhere along the caliper's range, and retention of the zero setting even when the caliper is turned off. With no moving parts in the readout, the digital caliper is exceptionally durable, standing up to some of the toughest manufacturing environments.

Concerns

Care and Respect. Like any measurement tool, the caliper must be treated with care and respect. Don't use it for purposes for which it was never intended (such as prying things apart). Wipe it clean after using, and don't throw it on the workbench. For dial calipers, be particularly wary of dirt which can accumulate on the rack, throwing measurements off and ultimately damaging the indicator. Store a caliper in its case. If it's going to be there for a while, apply a thin coat of oil to the jaws to inhibit corrosion.

Wear and Calibration. Check the caliper often for wear, as well as burrs and scratches on the jaws and contacting surfaces. A simple way to do this is to pass a master disc along the jaws while inspecting for wear or taper. Like any measurement tool, a caliper should be calibrated at least once a year or more often when use is heavy or there are multiple users of the same instrument.

Proper "Feel." While the caliper is a versatile tool, it is not one of the most precise. Skill is required for positioning the tool and interpreting the measurement result. As the user develops his "feel" for the tool, his measurement results become more consistent.

While the digital caliper may take some of the guesswork out of reading the measured value, it still requires skill on the part of the user to apply the tool properly to the dimension being measured. The jaws of the caliper must be square or perpendicular to the part. They are held firmly against the part, but not to the point of deflecting them. The part should be kept as close as possible to the frame of the measurement tool.

Knowing Its Limits. The rule of ten says that a measurement tool should have ten times more resolution than the tolerance of the dimension. Calipers typically read in 0.001" units. So if the tolerance is tighter than ±0.005", a micrometer (or some other higher accuracy tool) is the way to go.

The humble caliper is a surprisingly versatile tool for a wide range of general-purpose distance measurements. With a little skill, you can make a fast direct measurement or comparison in seconds and move on quickly to your next important task.

Micrometers: Measuring Under The Influence

The basic micrometer is one of the most popular and versatile precision hand-held measuring tools on the shop floor. While the most common type is the outside diameter style, the principle can be used for inside diameters, depths and grooves. With so many options for holding the spindle and alternate contact points available, it's a tool to satisfy an endless number of measurement applications.

The biggest problem with micrometers is that measurements are subject to variations from one operator to another. There are two types of influences that contribute to this variation: "feel" or inconsistent gaging force, and subjective factors.

The micrometer is a contact instrument. Sufficient torque must be applied to the micrometer to make good positive contact between the part and the instrument. The only torque calibration in the human hand is the operator's "feel." What feels like solid contact to one operator may not feel correct to another, so the readings will be different. In order to eliminate the "feel" part of the measurement, the designers of micrometers incorporated a ratchet or friction thimble mechanism. This is an attempt to assure more consistent contact pressure and eliminate the human influence.

A psychologist might say that the other type of measuring influences, the subjective ones, are all in the operators' heads. Tell an inspector that the best machinist in the plant made this part and influence enters the picture. Or suppose your boss walks over and asks you to measure a part and he adds, "I just made it myself." In these cases, measurements will tend to be better than the part deserves.

There are also more subtle types of influences. For example, if you know what size the parts ought to be before you measure them, readings will tend to be closer to that ideal than if the target dimension were unknown.

Don't take my word on this; conduct your own experiment.

Step 1: Take a number of workpieces and have several people measure them using micrometers without a clutch or ratchet or friction type thimble. Don't reveal the actual dimension of the workpiece or what anyone else got for readings. These are uninfluenced measurements.

Step 2: Give the same operators a known test piece to practice on to get a feel for obtaining a repeatable reading. Then ask them to measure an unknown part. Next, give another group a sample known part to practice on to get a feel for obtaining a repeatable reading; then have them measure parts where the size is known. These are influenced measurements, and I'm willing to bet good money that there will be significantly less variation in these results. It's just human nature.

Step 3: Replace the micrometer with one with a ratchet or friction thimble. The measurements are likely to improve even more.

Now that you have a better understanding of measuring under the influence you can do something about it. The simplest thing to do is use a hand tool that has ratchet or friction drives to achieve more consistent gaging pressure. Or, in the case of the micrometer, the best way to obtain the most consistent reading is with an indicating micrometer. This type of micrometer combines the flexibility of range with the high resolution and consistent gaging force of a dial indicator.

The lower anvil of an indicating micrometer is actually the sensitive contact of a built-in indicator which provides readings (it's typically in $1\mu m/50\mu"$ gradations) clearly and quickly with no vernier to read. Like the standard micrometer, you can adjust the spindle to the size needed and obtain a consistent gaging force when the master is set to zero on the dial indicator. Once established, the spindle is locked into position. Now the measuring tool begins to act like a gage by making measurements in a comparative mode. A retraction lever is also incorporated in the gage, making it easy to position the part for measurement quickly and to reduce wear on the contacts.

An indicating micrometer is a perfect gage for medium run, high tolerance parts. With this one gage an experienced operator can quickly set up the measurement process. Once the gage is locked in place, the indicating micrometer applies identical gaging pressure for each measurement, regardless of who is using it. The novice quickly obtains the same uniform high accuracy results as the experienced inspector regardless of differences in feel or what is known or not known about the part.

"Stylin" With Your Micrometer

Convenience is one of the reasons the micrometer is often the tool of choice for length/diameter measurements. The basic micrometer provides direct size information quickly, has high resolution, and is easily adaptable to many different measurement applications. Beyond the basics, there are all sorts of micrometer styles, which extend these advantages to many special measurement applications.

A micrometer consists of two opposing surfaces, a stationary anvil and a moveable spindle. On most micrometers, these hardened steel or carbide-tipped contact surfaces are flat.

However, micrometers can also be equipped with built-in or contact tips with unique forms for measuring special part characteristics.

Have A Ball. Ball contacts are used to measure wall thickness of tubes and other cylindrical components. Micrometers are available with one or two ball/radiused contacts. The one ball/radius style may be used for inspection of wall thickness on tubing. Two ball/radius contacts can inspect thickness between holes. In some cases the ball contacts can be supplied as attachments for use with a standard flat tipped micrometer. The attachments may be quickly and easily applied to either the anvil, the spindle or both. When using this type of attachment, the ball diameters must be taken into account by subtracting them from the micrometer reading.

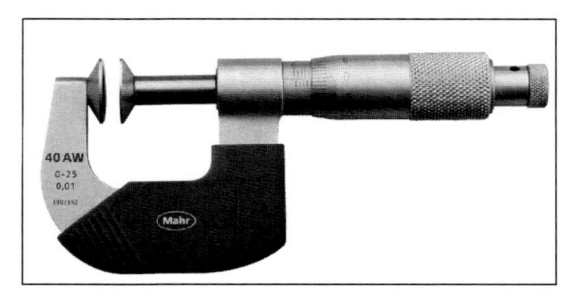

Time for Recess. Reduced spindle style micrometers have a turned down diameter on both the anvil and spindle. These contacts are used to measure inside recesses where the normal diameter may be too wide to penetrate. Because the contact areas of the anvil and spindle are very small, these micrometers may take a little getting used to. To get the proper "feel," take care to make sure the face of each contact is square with the axis of the diameter being measured.

In the Groove. Measuring the outside diameter of a cylindrical part from inside a turned groove on its surface calls for still another type of micrometer contact blades. Often these grooves can be so narrow that neither a standard nor reduced face micrometer will fit completely into the groove. Blade contacts, as the name implies, are very slender and flat. They nest readily into narrow-bottomed grooves. The blade solution created an interesting problem for the blade micrometer's designer. The spindle surface of most micrometers rotates as the micrometer barrel is turned, but a blade inside a groove would eventually be constrained from rotating. So blade micrometers have a spindle that slides along the axis of movement instead of rotating. Using this style micrometer calls for greater care. As always, check to make sure the micrometer is on the true diameter. Also, check frequently for wear on the measuring surfaces. Because the ends of the blades are so narrow, there is very little measuring surface. Excessive pressure on these narrow blades, as the tool is being rocked to find the true diameter, can result in premature wear.

Between The Grooves. On the same part, measuring the distance between the grooves is accomplished with a disk micrometer designed for thickness measurements on features that have narrow clearances. The measurement contacts are relatively large disk-like flats which extend beyond the diameter spindle and anvil. Because these contacts have such a broad measuring surface, parallelism errors can creep into the measurement. So it is important to check parallelism of the contacts using a precision ball on many locations between the contact faces. A discrepancy of more then a grad of the vernier is a sign that the parallelism of the anvil and spindle needs to be corrected.

Even the best and most basic hand measuring tool can be made better by adapting it to special application requirements. By choosing the most appropriate style of the application, you will achieve a faster and more accurate measurement. Each style, however, has its own unique requirements for care and use. If you're going to measure with style, make sure you know how to do it properly.

Stacking Up For Big ID's

For those medium and large parts with inside diameters greater than four inches, an inside micrometer is often used as the inspection tool of choice. This is especially true if the volume of parts is low and there is a large range of diameters to account for. Versatility of measurement range is one of the inside micrometer's most important characteristics.

This is actually one of the most straightforward of gages, since the gage itself duplicates the distance being measured. The axis of the inside micrometer *becomes* the diameter of the part. We mentioned Mr. Abbé a couple columns ago—you can't comply any better with his law than with an inside micrometer.

Usually in the world of dimensional measurement, it is frowned upon to add extension rods to a gage, since this can become a source of error. However, with the inside micrometer, this is exactly how the gage is used. In fact, the extension rods are all made to known reference lengths, as is the micrometer head. This way, the whole gage is put together with the extensions to become the size being measured. Though there is apt to be some error based on the accumulation of errors in the extensions, the large diameters themselves usually allow a larger tolerance, and typical gage tolerance rules can still be met. However, if a reference standard is available, it can also be used to convert the gage to a transfer-comparative gage for better performance.

The inside micrometer is similar to a standard micrometer, but without the frame. It is often sold in sets to allow for a wide range of diameters. Other characteristics include:

- A shortened spindle to allow access to smaller holes
- Spherical contacts which have radii smaller than the smallest radius they will contact
- Extensions that are manufactured to known lengths
- A collar to hold the extensions, which can be combined to reach diameters greater than 30".

However, while the inside micrometer does comply nicely with Abbé's Law, it is not without issues. The fact that it can be extremely long often makes it difficult to handle: sometimes it may even require two people. Plus, there is the fact that it's a two-point measurement without a third reference point. This means that it has to be rocked inside the diameter in two directions—axially and radially—searching for the maximum diameter. The best way to do this is to hold the reference contact against one side of the part and adjust it for fit, while moving the measuring end and simultaneously adjusting the micrometer for the best "fit."

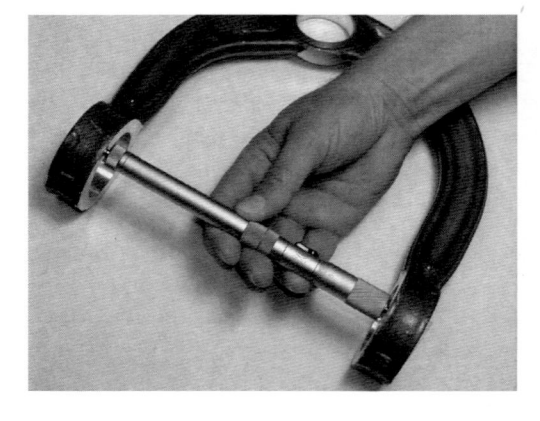

Temperature is the other issue to be concerned with. Since the gage has to be handled in order to be used, and the only way to handle it is to hold it, the gage is subject to the worst enemy of measurement: body heat transfer. This can be minimized though by using the gage for the shortest time possible, holding the gage only at the very extreme ends, and wearing insulating gloves. Some gages also employ insulated sleeves or holding areas to minimize body heat transfer.

While the mechanical inside micrometer version is still the most common set provided, technology has also stepped in to help the operator achieve better performance. By replacing the micrometer with a high precision, digital indicator with dynamic memory functions, operator influence can be minimized. With the gage in minimum memory mode, all the operator needs to do is sweep the ID, and the gage will search out the correct ID reading—lessening the need for rocking and the potential errors of trying to achieve the right feel. Though it is not without faults, the versatility and low cost of the inside micrometer gage makes it ideally suited for low volume part measurement applications.

■ ■ ■

Micrometer Accuracy: Drunken Threads And Slip-Sticks

With the number of micrometers on the shop floor and in inspection areas these days, it's important to understand the degree of confidence that one can expect from this instrument. Whether it's a screw thread or digital micrometer, the level of precision depends on two factors: the inherent accuracy of the reference (the screw thread or the digital scale) and process errors.

With a screw micrometer, accuracy relies on the lead of the screw built into the micrometer barrel. As with any screw-based movement, error in this type of micrometer tends to be cumulative and increases with the length of the spindle travel. This is one reason micrometers come in 1 in. (25 mm) measuring ranges. Apart from the difficulty of making long, fine threads, the error generated over the longer lengths may not be acceptable enough to meet performance requirements.

One approach that sometimes improves the overall performance of the measurement is to tune the micrometer to the range where it is most likely to be used. For example, if a 0 - 1 in. (0 - 25 mm) micrometer is to be used on parts towards the largest size, the micrometer could be calibrated and set up so that the optimum accuracy is at some other point in its travel than at its starting point. You could chose the middle to balance any errors at the end points, or elsewhere to maximize performance at any particular point of travel.

Aside from the calibration error of the thread, which reflects the accuracy of its movement per rotation, there are also two other thread related errors you should be aware of. One is error within the rotation, known as drunken thread, because of slight thread waver over the course of a rotation. The other is slip-stick, or backlash, which is caused by unwanted slop between the mesh of the threads. This is a common cause of reversal errors. As a point of reference, the drunken thread is like profile error on a machined surface, while slip-stick is similar to backlash errors seen in gears on dial indicators.

With electronic micrometers the thread usually drives a sensing head over a scale, or uses a rotary encoder as the displacement indicator. Both can induce slight errors, but the thread of the barrel remains the single largest source of error. However, an electronic micrometer can remember and correct for such errors, and in the end, provides better performance than the interpreted mechanical micrometer.

The process for checking the performance of a micrometer is similar to that of other comparative or scale based instruments. Gage blocks of known sizes are measured and the deviations from the expected values are plotted. Usually the gage blocks are chosen so that the spindle travels for a full or half turn of the screw. Taking this one step further, a single rotation of the screw can be analyzed by taking very small increments of measurements around the peaks discovered on the first pass. These small increments—maybe ten steps in one revolution—may reveal even larger errors, or show patterns that were machined into the screw threads.

Besides thread errors, the other significant cause for errors can be found in the parallelism of the anvils. The precision method for inspecting the condition of the anvils is with an optical flat. Using a monolithic light source, it is generally acceptable to allow two visible bands when assessing individual anvil flatness. For inspecting parallelism, a total of six bands may be observed, the combined total of both sides.

The applied measuring force of the sensing anvil on the part and the reference anvil is the other source of process measuring error. The friction of ratchet drive thimbles does reduce the deflection of the micrometer frame, but condition still exists as a source of error. With about two pounds of measuring force, typical frame deflection is roughly 50 microinches, although this is apt to increase on larger micrometers where the rigidity of the frame increases.

As with any hand tool measurement, other sources of errors will also sneak in. Temperature, dirt, and the means by which the operator aligns the gage to the part play a large role in the overall performance of any micrometer.

Evaluating Gaging For The Shop Floor
Understanding The New IP Standard

Measuring instruments have been used for the inspection of manufactured parts ever since the first vernier caliper was introduced. It didn't take much to take care of these old tools out on the shop floor: a clean cloth, a little elbow grease and a good storage box was all that was needed to make those gages last a lifetime. In fact, they often became prized possessions, as those old craftsman handed tools down from generation to generation.

In the past thirty years or so, electronic gages have become increasingly common on the shop floor because of their ease of use, speed, and ability to do complex measurements. However, when it came to caring for these new gages, one thing was clear: you didn't want to get that digital caliper, micrometer, indicator, amplifier or computer anywhere near water or coolant, or there was sure to be trouble. Either the gage wouldn't work, or even worse, it would produce incorrect readings.

This didn't make sense, of course, that the tools needed in an environment where there was coolant, grease, dirt and chips flying around did not like those conditions. Recently, there have been improvements in a lot of the electronic gaging that finally gives these tools the characteristics needed to survive out on the shop floor. Improvements in scale technology, microcircuits and sealing have made gages capable of literally making measurements under water.

Now that these types of gages are finally available, a new standard has been set up to help identify what type of tool is best for the environment in which it will be used. This rating is called Ingress Protection—IP for short. Associated with the IP is a two-digit rating number that tells what type of conditions that gage can survive in. The first digit describes the protection for solid foreign objects, while the second digit indicates protection against harmful ingress of water. A potential third digit, which is the defined impact protection, has not yet made its way into the measuring instrument table.

For example, a gage might have a rating of IP-65. As you can see from the accompanying table, this gage is totally protected against dust and protected against low-pressure jets of water from all directions, with limited ingress permitted. Today there are calipers and

First number (Protection against solid objects)	Definition	Second number (Protection against liquids)	Definition
0	No protection	0	No protection
1	Protected against solid objects over 50 mm (e.g., accidental touch by hands)	1	Protected against vertically falling drops of water
2	Protected against solid objects over 12 mm (e.g., fingers)	2	Protected against direct sprays up to 15° from the vertical
3	Protected against solid objects over 2.5 mm (e.g., tools and wires)	3	Protected against direct sprays up to 60° from the vertical
4	Protected against solid objects over 1 mm (e.g., tools, wires and small wires)	4	Protected against sprays from all directions - limited ingress permitted
5	Protected against dust - limited ingress (no harmful deposit)	5	Protected against low pressure jets of water from all directions - limited ingress permitted
6	Totally protected against dust	6	Protected against strong jets of water, e.g., for use on ship decks - limited ingress permitted
		7	Protected against the effects of temporary immersion between 15 cm and 1 m. Duration of test 30 minutes
		8	Protected against long periods of immersion under pressure

micrometers with ratings as high as IP-67. These can be subjected to the type of dust and dirt found in the shop and are both water and coolant proof.

So, electronic tools that can finally be used on the shop floor—what a good idea! But one cautionary note. Just because these new gages can handle the environment doesn't mean their measurements are impervious to environmental conditions. They are still precision gages and all the basic rules for precision gaging still apply.

Section N

Applications

Gaging The "Oddball" Application

The gages and methods discussed in this column are usually of very general interest. Virtually all metalworking shops need to measure holes, thicknesses, and heights. Some shops, however, have to perform measurements that are more limited in application: some are industry- or even company-specific.

Oftentimes, manufacturers attempt to satisfy unusual gaging applications by designing and fabricating a gage in-house. Most of the concepts of gage design are straightforward, and clever machinists can figure out how to apply these concepts in fairly basic fashion, usually by incorporating stock components such as dial indicators and gage heads, into custom-made fixtures. Amateur gage design, however, may not be efficient. It's often more cost effective to let gaging experts work out the details and spread design costs over a larger number of users.

Following are examples of gages developed for specialized applications. There are two points to be made here: the first is that even the most obscure measurement tasks can usually be performed by relatively simple gages. The second is that many gage makers are more than willing to work on "oddball" applications, freeing gage users to concentrate on their real business.

There are hundreds of thousands of countersunk rivet holes in the skin of a typical airliner. To achieve maximum holding power and minimum wind resistance, the head of every rivet must sit flush with the outer surface of the skin. However, government standards strictly limit the degree to which a manufacturer can grind down proud-standing rivet heads. As countersink bits wear, countersink diameters and depths change. Riveters need to measure the countersinks, in order to pick the right size rivet from an available selection range.

Countersink gages operate like hand-held, plunger-type depth gages with special-ratio dials or electronics, to convert vertical movement at the contact into diametric measurements. The gage contact must be chosen to correspond with the included angle of the countersink. Some gages measure the major, or entry diameter, while others measure the depth of the minor diameter (where the countersink meets the straight hole). Although aerospace is currently the predominant application, other high-precision sheetmetal industries may soon follow suit.

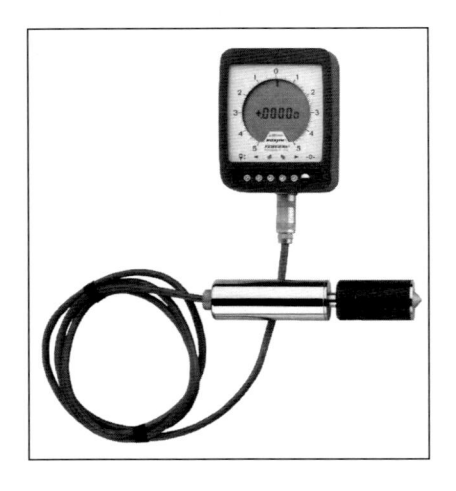

More exotic is the Almen gage, which is used to measure and control the results of the shot peening process. Shot peening bombards a part with a stream of steel or glass particles, inducing sub-surface material compression and greatly reducing stress cracking under tensile loads. Shot peening is widely used on automotive and compressor crankshafts and conrods.

To measure the process, a thin metal test strip is subjected to the shot stream. The strip bends in a predictable manner, as a function of shot stream intensity. An Almen gage measures the radius of the strip. Gage manufacturers have developed innovative mastering strategies to take both the longitudinal and transverse curvatures of the strip into account, and use refined mechanisms and geometries to hold and measure the strip without distortion. Although the Almen gage is essentially a benchtop depth gage, it's these important details that make the gaging process quick and cost effective.

Aerosol and beverage can manufacturers have highly specialized gaging requirements. Tight tolerances must be maintained, both in order to ensure proper functioning of valves, seals, and pop-tops, and to save material: at production rates of several million units per day, a wall thickness variation of a few millionths quickly adds up to tons of aluminum. There are more than a dozen critical dimensions on most cans, including such arcane features as crimp groove location, valve stem height, and "tab top bubble height."

In order for SPC to be effective—in order for gaging results to work their way back to the manufacturing process in a timely manner as the cans go whizzing by—several dimensions must be measured very rapidly. Whole catalogs of can gages have been developed. At the lower end of the technology spectrum are specialized indicator gages, which differ from generic gages mainly in the shapes of their reference surfaces and sensitive contacts. Once again, the dimensions being measured here are simply variations on the basic themes of height, depth, thickness, ID, and OD. But the theme can be elaborated almost infinitely, depending upon the need for precision, throughput, and electronic output. Engineered fixture gages incorporating air jets and/or electronic gage heads can get the manufacturer halfway there, while fully automated systems with parts handling capabilities can ultimately make can gaging a hands-off, closed-loop process.

If the only concern is accuracy, then gage design can be pretty simple: laboratories achieve extremely good results with a dial indicator and a comparator stand. Production applications, however, impose additional demands of throughput, ease of use, delivery, and maintenance and mastering intervals. When it comes to cost effectiveness in specialized applications, it's the details that make the difference.

■ ■ ■

Can't Measure It? Try A Caliper Gage

If a workpiece has ever left your shop with an important dimension unmeasured because you couldn't get at it to measure it, you probably didn't know about indicating caliper gages. Regardless of the shape of the housing or the complexity of the casting, no matter what size the flange, what curve the tubing, what width the material, or what depth the recess, an indicating caliper gage can give you the desired thickness or inside dimension. Non-destructively, let me hasten to add.

It is important to distinguish indicating caliper gages from the familiar 0"- 6" calipers.

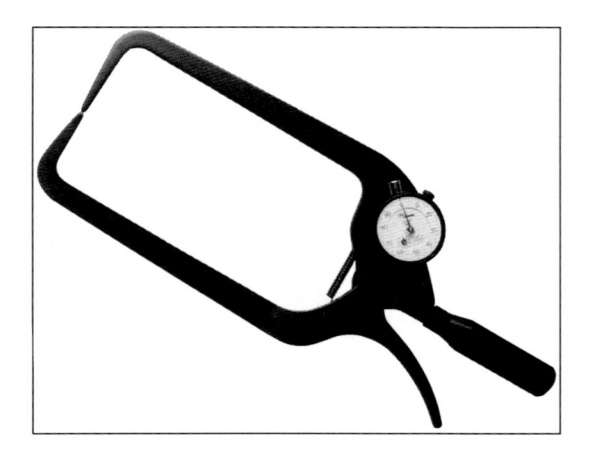

These latter, though useful tools, are only capable of measuring the basics—IDs, ODs, lengths and depths. Indicating caliper gages, on the other hand, use dial indicators. Most have resolutions of .010", and some go to .001". With a range from 0" - 1", or 0" - 3", the indicating caliper gage incorporates a scissors action to facilitate getting around obstructions, and is an immensely useful tool.

The key to the caliper gage's flexibility is its inherently simple hinged geometry: movement at the contacts is mechanically reduced by a gear at the pivot, then re-enlarged the same amount on the face of the indicator. (The ratio is usually 10:1.) As long as that ratio is maintained, the jaws can be essentially any shape you want. They can curve over and around any type of flange, into any curved or angled hole, across long distances, and into the most inaccessible recesses.

The thing to remember about caliper gages is, if you don't see it, ask for it. Our catalog, for example, lists eight standard gages, but we have engineered—and I'm not exaggerating—over 50,000 "specials" for customers ranging all over the board. To design a special caliper gage all a manufacturer needs is a print of the part to be measured, although a sample can also be helpful.

Besides jaw shape, other common options include jaw size (up to 4' long!), body material (aluminum, magnesium, honeycomb composites, etc.), contact shape (ball, blade, rollers), and contact material (carbide, ruby, diamond, plastic). Of course, the indicators come in inches or metric, and any kind of custom dial face can be designed to suit the application. Revolution counters are standard to help you keep track of large dimensions. Because they're fairly simple tools, prices are moderate: most specials cost $700-$800, although prices as low as $500 and as high as $5,000 are not unheard of.

Where would you use an indicating caliper gage? Just a few applications that spring to mind are: housings and castings of all kinds; valve bodies; manifolds; tubing; aircraft components (turbine blades, body and wing components, fuel tanks); wide sheet-type materials (steel, laminates, plywood, composition boards); air-cooled engine cylinder castings (check wall thickness between the fins); and toilets. I know a jet engine manufacturer who has a dedicated caliper gage for every critical dimension on his engine housing castings: literally thousands of gages.

Outside caliper gages measure outside dimensions, such as cylinder or tubing wall thicknesses, flanges, and sheet-stock. Inside caliper gages are often the only way to check inside dimensional features like seal and bearing seats deep inside a casting, recess IDs, or IDs of bent tubing.

Even if your tolerances are relatively coarse fractional numbers that you could measure with a pocket scale or a 0" - 6" dial caliper, or if you currently use a Go/No Go gage, an indicating caliper gage can provide benefits in a production environment. Rather than waiting until you bump up against the next tick on the scale that says you're at your tolerance limit, or worse, make a bad, "No Go" part, the higher resolution and magnification of an indicating caliper allows you to view the trend of your process. You can see if your dimension is getting larger or smaller, be it ever so gradually. This permits you to make adjustments so that you stay near the center of your tolerance range, and avoid the limits altogether.

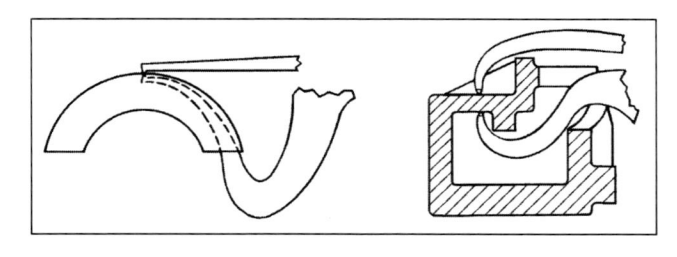

In sum: Within the limits of their ability to resolve, there's virtually nothing you can't measure with an indicating caliper gage.

The Versatile Surface Plate

Surface plates provide a broad, smooth, flat reference surface that can be extremely useful for inspecting incoming, in-process, or finished parts. When used in combination with various gages and accessories, such as height gages, gage blocks, angle plates, and squares, they can be used to check a wide range of parameters, including length, flatness, squareness, straightness, angle, feature location, and runout. Surface plates are simple, and extremely versatile.

Surface plates come in a wide range of sizes, from about 12" × 12" to 6' × 12', and weighing up to 10 tons. Three grades are available, the flatness tolerances for each grade varying with the size of the plate: AA (laboratory grade); A (inspection grade); and B (toolroom grade). Many can be ordered with ledges and threaded inserts, both of which make it easier to clamp workpieces or accessories to the surface. Granite is the most common material used: it is harder and denser than steel, has very little internal stress, and is less subject to dimensional change due to temperature variations.

Care and maintenance is basic, but important. A surface plate is a piece of precision gaging equipment, not a storage table. Use it for measurement purposes only, and keep it clear and covered when not in use. Before use, wipe it down with a special surface plate cleaning solution. Make sure that workpieces and gage bases are clean and free of burrs be-

Section N: Applications

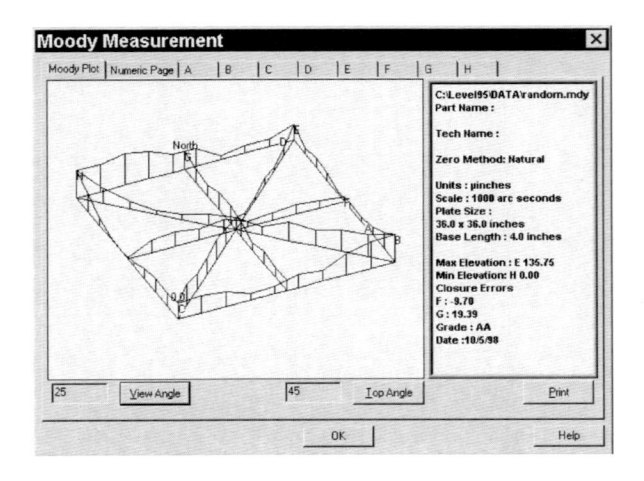

fore placing them on the plate. Don't drop anything on it.

Many uses of the surface plate are extremely simple. Some milled or ground parts can be checked for flatness by placing them on the plate and moving them around to see if any rocking motion is possible. If so, feeler stock can be used to find the high spots and measure the variation.

Most often, surface plates are used in conjunction with a portable height stand, supporting a dial or test indicator, or an electronic lever-type gage head. The height stand has a fine-ground base, which allows it to be slid across the surface without scratching. For a simple flatness or parallelism check, the gage is brought into contact with the part and zeroed out. Then the gage is moved around on the plate to "explore" the part surface for deviation. Since the surface plate is the reference, deviation may be errors of flatness, parallelism, or both.

The same setup can be used to check heights, using gage blocks as the height standard. A more sophisticated approach is to use a height master, which combines a permanent stack of blocks, staggered left and right, with a micrometer height adjustment: this allows the user to set both top and bottom heights anywhere from 0" to 12" in .0001" or .00001" increments.

"Smart" height stands can be used to substantially automate measurements on surface plates. Sometimes called "single-axis CMMs," smart height stands consist of a vertical slide with a position encoder, a lever-type electronic gage head and/or touch probe, and an electronic keypad control. They can be programmed to measure many part features and dimensions, including diameters, lengths, and locations, in any desired sequence.

Aside from the measuring devices themselves, a number of positioning accessories extend the usefulness of surface plates. Vee blocks serve as simple holding devices for cylindrical parts. Placed on their sides, they become a clamping surface for shafts, in order to measure the flatness or squareness of the shaft ends. Vee blocks are particularly useful for measuring runout. A part is placed in the vee and rotated, while an indicator or gage head measures the variation in height. (Note that lobes on a round part may create an unstable axis of rotation

If projecting features prevent a part from being placed flat on a surface plate, a parallel bar accessory may be used to bridge across those projections. Sine bars and sine plates are used to establish surfaces at precise angles from horizontal. Angled parts are placed on the angled surface, to check them for flatness, or to measure the angular accuracy of the machined surface.

The surface plate itself can be a gage. A hole is bored in the plate, and a gage head or air jet installed, to inspect flush surfaces for flatness without the use of a height stand. If desired, a second gage head in a height stand can be positioned directly above the one embedded in the surface plate, permitting independent measurements of flatness, thickness, and parallelism.

Surface plates provide a stable reference surface on a large scale, making a great many gaging setups possible on a single, simple piece of equipment. When a gaging application does not warrant the purchase of a special-purpose fixture gage, surface plates often provide an economical, all-purpose solution.

■ ■ ■

The Plane Truth About Flatness

The flatness of machined planar surfaces is often critical to the performance of parts and assemblies. The plane is also the basis, or reference, for most dimensional and geometric measurements, including height, location of features, squareness, and datums. A reference plane may be a feature on the part itself, or it may be part of the measuring instrument, but in either case, the measurement can only be as accurate as the reference. So whether you're making parts or measuring them, you may have to measure flatness.

There are many tools and methods available, depending upon the nature of the part and the degree of accuracy required. Surface plates serve as a general-purpose reference for many flatness measurements. If the flat surface of the workpiece can be put in direct contact with the plate, it is possible to measure flatness using feeler stock, although this is a low-resolution method, and only the perimeter of the part is accessible. An air or electronic gaging probe installed flush in the surface plate can provide much higher resolution, if the part is small enough to move around on the plate. Each type of probe has its benefits. Air jets are self-cleaning and non-contact, while electronic transducers can be connected with gaging amplifiers or remote indicators with dynamic measuring capabilities, to automatically capture the maximum deviation, or to output data for SPC.

If the part is too big to slide around, or if its configuration is such that the flat surface can't be put in direct contact with the surface plate, then it must be staged. A test stand with a mechanical indicator or an electronic gage head is slid around on the surface plate to explore the part.

This, however, may fail to distinguish between errors of flatness and errors of parallelism. To break out flatness, measurements are taken at equally spaced points on the surface, then the data is plotted on a graph and a best-fit line calculated. Deviations from the best-fit line represent errors of flatness. If the measurements are taken on a vertical surface (using, for example, a "smart" height gage with the gage head turned 90 degrees), one would duplicate the procedure to break flatness out from possible squareness errors.

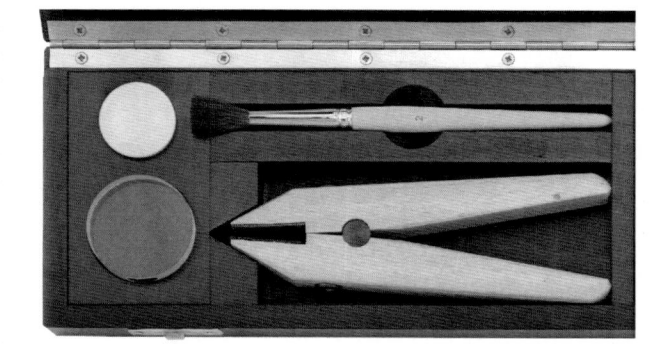

To measure really large areas, like machine beds or surface plates, electronic levels are often the appropriate tool. Levels may be connected to gaging amplifiers that will automatically convert angular readings into dimensional error. Large areas can also be measured with electronic probes, using a precision straightedge as the reference, as described previously.

With the proper software, the data obtained from large-area flatness measurements can be converted into a 3D plot. This information can be used in at least three ways: the user can do his setups on the flattest areas and avoid the worst sections of the surface plate or machine tool; he can use the data to compensate mathematically for out-of-flatness; and he can use it as a guide to correct the out-of-flat condition.

Optical flats are references for measuring small, high-precision parts, such as gage blocks. Usually made from fused quartz or high quality glass, the puck-shaped optical flat is certified to within 1, 2, 4, or 8 microinches. It is wrung to the part and viewed under a monochromatic (helium) light source. A perfectly flat part will reflect straight, regularly spaced, easily visible interference bands, each representing an interval of 11.6 microinches (the half-wavelength of helium light). Air gaps (i.e., low spots) between the part and the flat will distort the interference bands proportionally to the flatness error: a band that is "bent" by one half its thickness indicates out-of-flatness of 5.3 microinches ($^1/_2 \times 11.6$). The location of low spots can be identified by the direction of the distortion.

Regardless of the method, before a part can be measured for flatness it is important to know the level of uncertainty in the reference. Flatness may be transferred from certified standards to masters, then from masters to gages, and thence to workpieces, but be aware that the level of uncertainty increases at each step.

Bar Talk: What's Your Sine?

Previously, this column described the use of surface plates, observing that a flat surface is the basis for most dimensional measurements. Many workpieces, of course, are neither flat nor straight. In order to measure the angular accuracy or straightness of an angled surface using a surface plate, a sine instrument comes into play. By placing the workpiece on the sine instrument and raising one end of the instrument to the proper height, it is possible to orient the workpiece parallel to the surface plate. Straightforward measurements can then be performed with a test indicator and height stand.

Sine bars are relatively narrow (up to about 1" wide) fine-ground or lapped steel bars, with precision cylinders at each end resting against stops machined into the bottom surface. Some sine bars are completely ambidextrous; others have "upper" and "lower" ends as well as distinct top and bottom surfaces. Holes machined in the instrument enable the placement of stops or clamps to hold workpieces in place. Close cousins to sine bars include sine blocks, which are simply wider sine bars; and sine plates (the most popular version), in which the bottom cylinder is actually part of a hinge connected to an attached base. While most sine instruments serve primarily as measuring instruments, some sine plates are rugged enough to serve as fixturing devices for machining operations. The principle of operation is identical for all versions, so for the sake of simplicity, we'll refer only to sine plates in the following discussion.

To set the sine plate at the proper angle, one simply selects a gage block or gage block stack of the appropriate height and places it under the upper cylinder. An imaginary right triangular prism is thus created, the vertical face of which passes through the gage block and terminates at the axis of the upper cylinder. The horizontal base of the prism is above and parallel to the surface plate, terminating at the axis of the lower cylinder (hinge), while the ends of the hypotenuse are defined by the axes of the two cylinders. The top surface of the sine plate is therefore parallel to the hypotenuse. The use of cylinders as contact points ensures that the length of the hypotenuse remains the same regardless of its angle. Most sine instruments set the cylinders at a fixed distance between centers that is easy to manipulate mathematically, 5" and 10" being the most common lengths.

To calculate the required height of gage blocks, use the following formula:

$\sin B = b/a$ where:

B = required angle
b = height of triangle
a = length of hypotenuse.

For example, say we want to measure the straightness of a surface on a workpiece that's angled 27° from a reference surface on the same workpiece. Let's also say our sine plate is 5" between cylinder centers. To obtain the sine value, we either refer to a table of trig values, or simply punch it up on a $10 scientific calculator.

$\sin 27° = b/5"$
$0.453,990,499 = b/5"$
$b = 2.270"$

So we wring up a stack of gage blocks, place it beneath the upper cylinder of the sine plate, and *voilá*, we have a surface that's 27° out-of-parallel with the surface plate. We stage the workpiece on the sine plate facing the opposite direction and, if all is well, the workpiece surface should be parallel to the surface plate. We can then measure it for flatness, straightness, and angular accuracy, using conventional surface plate methods. To measure straightness, an indicator held by a test stand is drawn across the surface, and comparative height measurements are taken at regular intervals. By drawing a best-fit line through the data, it is possible to break out the flatness of the surface and to calculate the accuracy of the machined angle.

Because sine values change rapidly between angular measurements of low numerical values, and change very little between higher numerical value angles, the accuracy that can be achieved using a sine instrument varies considerably: precision is much better for shallow angles than for steep ones. Therefore, where the surface of a workpiece is angled at greater than 45°, it is often advisable to use the complement of the angle (90° - x) in the calculation.

Even so, the accuracy that can be achieved with sine instruments is somewhat limited, because there are so many separate mechanical elements to the set-up (the surface plate, the gage block(s), the test indicator, the test stand, and the sine instrument itself), each of which imposes a certain degree of uncertainty. We'll look at methods to perform higher-precision angular measurements in another column.

Measuring Tapered Parts

Recently, we looked at surface plate methods to measure angles, involving the use of sine plates and indicator stands. By mounting a V-block on the sine plate, these methods can be used to measure tapers on parts such as shafts or conical workpieces. This kind of painstaking layout work, however, is not appropriate for the high volume, tight-tolerance requirements of machine shops. For production-environment measurements of tapered features, other methods are needed.

Common precision tapered male and female parts include features on engine shafts, spool valves, fuel injectors, and needle jets. Among the highest-precision applications are toolholder tapers, and components for orthopedic artificial joints, where tolerances may be in the range of 1-2 arc-sec. or 5-10 microinches per inch. The angular conformance of mating tapered parts can have a major influence on the performance of these assemblies.

As with most dimensional measurements, tapers can be measured with a variety of methods and hardware, offering a wide range of pricing, accuracy, speed, required user skill, numerical capabilities, and other characteristics. We'll start at the low end of the price and accuracy ranges and work our way up.

If numeric accuracy is unimportant, and a general sense of conformance is adequate, one might use a male or female master taper that mates with a workpiece of the opposite gender. This is not a real form of measurement, *per se*—it's more a form of go/no-go gaging. The master and workpiece are simply mated, and conformance is determined by feel. If bluing dye is placed on the mating surface of the master, it will transfer to the workpiece wherever the two come in contact. This will show whether there is too much or too little taper, and give an indication of out-of-roundness and some other geometric errors, but it provides no numeric results: results are entirely subjective. The precision of this method is quite low—probably on the order of one or two thousandths of an inch—so it's really only good for situations when "close enough" really is close enough.

Air tooling and fixture gaging incorporating either electronic or air probes, represent a significant step up in capabilities. These are generally custom-built devices, but need not be expensive, for most of them are quite straightforward. Probes or air jets are arranged to measure two or more diameters on the tapered feature at a known distance or height from one another. Female tapers can

Female tapers can be measured using tapered air plugs, while male tapers that appear at the end of a workpiece can be measured with tapered air rings.

be measured using tapered air plugs, while male tapers that appear at the end of a workpiece can be measured with tapered air rings.

Sometimes it is lack of taper that is important. Crankshaft journals, for example, must be checked to ensure there is no taper, barrel, or hourglass shape. This is best performed with a fixture gage, which holds the shaft in V-blocks or centers, while two or more pairs of electronic or air probes measure the journal from opposite sides. Mechanical indicators can also be used, but this is becoming less common as the requirements for speed and high precision increase. Today's amplifiers and air systems, with their computing and output capabilities, have all but replaced the mechanical indicator for these fixture-gaging applications.

Normally, this methodology involves comparative measurements, in which an accurately machined taper generates a reading of zero on the gage, and deviation from nominal is shown as the sum of the deviation, in opposite directions, of the two measured diameters. In other words, if the larger diameter, Diameter 1, is 0.0001" over nominal, and the smaller diameter, Diameter 2, is 0.0002" under nominal, the gage will read total deviation as +0.0003". On the other hand, if Diameter 1 is 0.0001" over and Diameter 2 is also 0.0001" over, then total deviation in the taper is zero: there may be a machining problem in holding the diameters to spec, but the taper itself is accurate. Many gaging amplifiers can readily convert gaging input into absolute, as opposed to comparative readings, and present the results in units of arc-sec., or microinches of error per inch of tapered height.

Comparative-measurement tooling and fixture-type gaging represent the best methods for use in production environments, in terms of price, throughput, ruggedness, and environmental stability. Very little operator skill may be required, depending upon details of the gaging amplifier or other input/output options.

Naturally, these methods require very accurate masters. In order to check the masters, an even higher level of accuracy is required, which can be achieved with a master-ring-and-disc comparator. This is similar in principle to a typical benchtop ID/OD gage, but raised to a very high level of technical refinement, with precision to 1 microinch possible in a controlled environment. By raising or lowering the instrument's jaws, diameters can be checked at two or more heights; then a simple trig function is applied to calculate the taper. This type of gage, which is normally confined to laboratory use, tends to be on the pricey side, and is engineered for accuracy rather than throughput as the top priority.

■ ■ ■

Depth Gages

Depth gages are among the simplest of indicator gages, typically consisting of an indicating device mounted through a reference bar or plate. Though they may be simple, depth gages are used in thousands of critical applications, to measure the depth of holes, counterbores, slots, and recesses, as well as heights or locations of some features.

The first depth gages consisted of a simple rule with a sliding perpendicular beam as the reference. As the needs for higher resolution and precision increased, these were largely replaced by vernier devices and micrometer depth gages. And while both verniers and micrometers remain in wide use, indicator depth gages provide even higher levels of accuracy, as well as increased speed of operation and lower dependence upon operator skill.

As with almost all indicator gages, depth gages can be readily modified to suit particular application needs, especially to make high-volume gaging tasks quicker. Depth gages are available with various styles of indicators, contact points, and bases.

The simplest and most common depth gage has a flat base or anvil, a sensitive contact that retracts flush with the base, and a radiused contact point. This is an absolute gage, measuring the full depth of a feature, from zero out to the indicator's maximum range. No master is required: to zero the gage simply set the base on a precision flat surface.

Different contacts can be used to tailor the gage to special applications. For example, by replacing the standard radiused contact with a needle-style contact, it is possible to measure surface pits, small holes and recesses, and etch depths.

Extended contact points can be added to measure greater depths, or to turn an absolute-measuring gage into a comparative gage. Such a gage can be mastered with gage blocks by holding one end of the base firmly on top of the stack, with the spindle as close to the stack as possible without interference. Special depth masters, however, are quicker and more reliable, and are thus more practical for production gaging applications.

Special bases can also increase gaging efficiency. Counterbores may be gaged more easily if the indicator is offset from the centerline of the base. V-shaped bases are useful in applications where a standard flat base would interfere with the user's ability to locate a needle-type contact in a small feature, such as a pit or an etched line. The V-base provides a wider viewing angle, but still has a narrow "flat" on the bottom to help orient the gage perpendicular to the part surface. The user first tips the V-base on the workpiece surface, locates the contact point in the feature, then "rolls" the gage upright until it rests on its flat.

Custom anvils can be readily designed to conform to the shape of the workpiece. Take, for example, the aerosol can. This is a metal part that is literally under pressure, and so is liable to more potential failures than most types of containers. The depth of the crimp groove is a critical quality dimension that must be carefully monitored. Depth gages designed for this application with special bases that rest securely on top of the can have proven themselves ten times faster in use than generic vernier depth gages.

All of the gages described above are portable or hand-held designs, which implies bringing the gage to the workpiece. It is often convenient, however, to bring the part to the gage, especially if the part is small. Benchtop depth gages essentially turn the portable gage upside-down, and provide a wide flat reference surface—virtually a table—upon which the workpiece can be placed and manipulated. Parts can also be "explored" for flatness with this type of gage, by sliding the workpiece around on the table.

Users can also choose among indicator styles. Long-range indicators, with revolution counters, can measure depths from 0" to several inches (or their metric equivalents). Special indicator faces can be designed for "stoplight" gaging, with green, yellow, and red segments to quickly signal good, marginal, and out-of-tolerance parts. Indicators with "push-down" movements allow users to locate the contact point against the workpiece more positively than is possible with conventional "sprung-down" indicators. Gages can also be equipped with digital electronic indicators, providing opportunities for dynamic measurements (such as automatic capture of minimum or maximum readings), and data output.

Rules for depth gage use are straightforward. Both workpiece and anvil must be clean and free of burrs. For portable gages, the base must be held firmly against the workpiece, and it must be positioned flat and square. As with all indicator gages, accuracy also requires a rational mastering schedule, the frequency of which depends upon the amount of use, as well as the conditions in the gaging environment.

■ ■ ■

Got A Match?

Producing precision spools and sleeves (e.g., for fluid pumps) and other pairs of parts with matching inside and outside diameters can be among the trickiest of quality issues. It's very easy for an engineer to specify that the outside diameter (OD) of the one and the inside diameter (ID) of the other must be within .000025"—and less easy for a machinist to accomplish.

You can spend days trying to tighten up your processes to meet those specs consistently, and chances are you'll still end up tossing or reworking a high percentage of your production. The solution to the problem may lie not in the process, but rather, at the QC end. Maybe you can loosen up on your process, and measure your way out of the problem with match gaging.

When an engineer specifies tolerances for spool and sleeve diameters, what is his real concern? Is it absolute dimension? Likely not. In an application like a fluid pump, or a fuel injector, it's the clearance between the two parts that determines how well it functions. That's where match gaging comes in. Match gaging doesn't measure diameters: it measures the clearance (or interference) between two parts. It can be a tremendous time- and work-saver where a desired amount of clearance or interference is required.

In its simplest form, match gaging uses an air gage with a two-legged manifold, one leading to an air plug, the other to an air ring. To measure a match, place the OD part (spool) in the

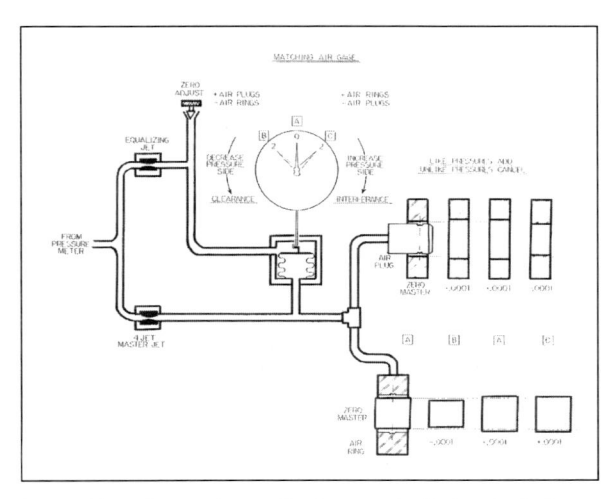

A matching air gage is an effective way to find parts that meet specified clearances between them.

Section N: Applications   219

ring and the ID part (sleeve) on the plug. The gage indicates the total of the clearances between the two parts and their respective fixtures.

Match gages are available in a wide range of configurations from simple manual gages to highly complex, fully automatic, multi-measurement gages costing many thousands of dollars. In terms of accuracy, they are available with 50 millionths resolution and a range of 0.003" at the coarsest, to 5 millionths resolution and a range of 0.0003" at the finest.

The greatest advantage of match gaging is that it allows you to produce matching parts at extremely tight clearance tolerances without actually having to achieve the same level of precision in the machining process itself. To see how that works, let's look at a set of parts with a nominal size of 1.0000" and where engineering has calculated that optimal performance requires a clearance tolerance of between 100 microinches and 200 microinches.

There are basically three ways of achieving this desired fit between matched components. The first involves controlling the size of both parts to assure an accurate match and full interchangeability. This method tightens the manufacturing process severely and necessitates machining to a plus tolerance on the OD part and a minus tolerance on the ID part, as follows:

Outside Diameter   **Size Tolerance**
1.000100"   +0.000050"/-0.000000"
Inside Diameter   **Size Tolerance**
1.000000"   +0.000000"/-0.000050"

The second method is to control the size of one component where the other is predetermined by a previous process. Typically, the OD is the fixed measurement (in this case, 1.0000" ±0.0001") and the ID is then machined to fit. Air gaging is used to assure the 100μ inch to 200μ inch clearance between parts. This method usually requires maintaining a substantial inventory of OD parts and often involves a cumbersome measuring process.

Match gaging gives us a third alternative: we can machine both components, allowing the tolerances to vary as they will, and use the match gage to select matched sets. Thus, the OD is machined to 1.0000" ±0.0005" and the ID to 1.0000" ±0.0005". Assuming a normal distribution curve from our machining process, we can use a match gage (often automatic) to select sets, which fall within our 100 microinch tolerance range.

Some users may elaborate on this process. For example, a manufacturer of fuel injectors had a clearance tolerance range of 20 microinches between barrel and spool. They were unable to control the machining process to produce either part consistently within that range, but a fully automatic match gaging setup now allows them to stage, match, assemble and package 700 pairs per hour. Feedback from the gage is used to control the processes, letting the distribution of spool ODs move up or down to accommodate an overabundance of barrel IDs at one end of the scale or the other. Both ranges are allowed to float, chasing each other up and down the scale to assure enough matching spools and barrels to maintain production rates.

In The Groove

It just might be easier to manufacture a groove on a turned part than it is to inspect it. Select the right cutting tool, set the CNC, and *bada-bing bada-boom*! You're done. (Okay, I exaggerate.) But because of the critical functional role played by grooves for seal rings and retainer rings, good gaging practice is a must. There are many different gage types from which to choose, and because inside grooves tend to be more difficult to measure than outside grooves, we'll examine inside groove gages more closely.

The pistol-grip or linear-retraction style groove gage is specifically designed for inside grooves. It features long retraction of $1/2$ inch (13 mm), which enables it to fit inside the bore, then expand to the full depth of the groove. The indicator's sensitive range is usually much smaller than the retraction—on the order of 0.040 inch (1 mm), which is more than adequate for the expected variation in most precision applications. The indicator may be either dial or digital, and usually has resolution of 0.0005 inch (0.01 mm). The lower contact is a fixed or reference contact, to bear the weight of the gage; the sensitive contact is located above, where it is unaffected by the weight.

Contacts tend to be built robustly, to resist flexing when the gage is "rocked," in the same manner as other bore gages. Procedures for mastering, too, are similar to those used for bore gages, except that a fine adjust is usually provided on groove gages, to simplify zeroing.

Contacts are adjustable over a range of 2" to 4" (50 mm to 100 mm), so that different size parts may be inspected with a single gage. Another important characteristic is the interchangeability of contact tips. Extra-long contacts permit access to deeply spaced grooves, and special contact shapes allow inspection of grooves right up to the shoulder, and of round- and V-bottom grooves.

Inside-measuring swing-arm or caliper gages are general-purpose instruments that may be applied to a wide range of inside dimensions, including inside grooves. Caliper gages can measure through their entire retraction range, which is often over $3/4$ inch (19 mm.) These gages can therefore measure features where variation is expected to be large, and can measure different size parts without adjustment. On the other hand, because they are non-adjustable, their ultimate range and flexibility may be poorer than pistol-grip style gages.

Because of their long range, caliper gages have dial indicators with secondary dials, or revolution counters. Some users have trouble reading these dials accurately. One way to bypass this problem is with a digital caliper gage, although these are not very common.

Rocking a caliper gage side-to-side for axial alignment tends to stress the caliper arms and pivot, and is not recommended. As a result, caliper gages may not produce results

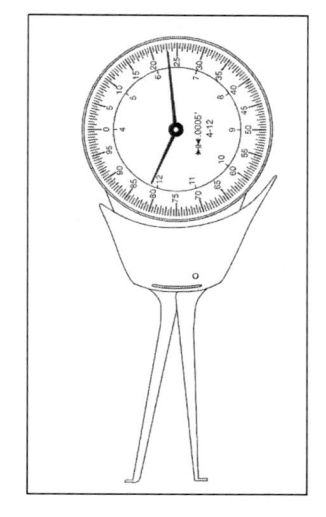

When choosing a caliper-style gage for inside-diameter groove measurements, users must weigh issues of range, capacity, ruggedness, accuracy and cost.

Section N: Applications 221

as accurate as pistol-grip gages, which are engineered to promote rocking. On the other hand, caliper gages are lighter, smaller, less expensive, and easier to use, all of which makes them quite popular. They also offer a greater variety of contact styles for access to difficult features.

Compared to inside grooves, outside grooves are a snap. They are normally measured with a snap gage fitted with special narrow anvils, called blade anvils, which are available to match the depth, width, or shape of any groove. With their heavy spring pressure and rugged positioning backstops, snap gages do not need to be rocked, and are very easy to use.

Finally, bench-type ID/OD comparators can measure inside and outside grooves if fitted with special contacts. ID/OD comparators offer the highest potential accuracy, and allow the part to be rotated, to explore the groove for minimum or maximum readings. But they are limited to use with small, portable parts, with grooves less than 1"/25 mm from the mouth of the bore.

■ ■ ■

"Squeeze" Gaging

Machinists working strictly in metalworking shops do not have many occasions to gage the thickness of soft materials. But many of our readers work in supporting roles, helping to build or maintain the machines that produce textiles, plastic films, paper and other products that are compressible. Even readers who work only with metals should realize they are not alone in their concern for accuracy: the shirt on your back, the newspaper you read this morning and the garbage bag you tossed it in, were all produced to exacting thickness tolerances.

Differences in gaging pressure translate readily into differences in the degree to which the sensitive contact compresses or penetrates the material being gaged. It is, therefore, essential to use a standardized method when measuring the thickness of compressible materials. Every industry and every kind of material has its own standards, most of which have been established by the American Society for Testing Materials or by Underwriters' Laboratories. These standards specify the size of the reference table and the measuring contact (or foot), the dial gradations, and the amount of force

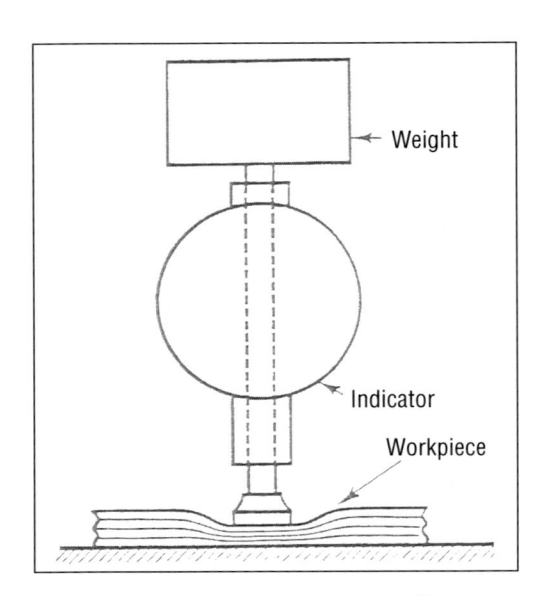

When gaging compressible materials, differences in gaging pressure translates readily into differences in the degree to which the sensitive contact compresses the material being gaged. To ensure consistency of compression, gaging force is controlled by a weight, rather than with the springs found in most dial indicators.

to be used. (Gaging force is controlled by a weight, rather than with the springs found in most dial indicators, to ensure consistency of compression). Such standards make purchasing decisions quite straightforward: often, you need only specify which ASTM standard you wish to meet.

Assuming you have the right gage for your material, making accurate measurements of compressible materials is not difficult if you follow a few basic guidelines:

The Anvils Must Be Parallel

If they are not, they will compress the material at an edge rather than across a flat surface. To check for parallelism, close the contacts and try to shine a flashlight through them. Parallel contacts allow no light to shine through; any lack of parallelism is readily apparent down to about 0.0001". Check front to back, and side to side.

It is difficult to hold larger contacts—say those above $1/2" \times 1/2"$—in alignment with the light source. And the visual method may fail to reveal problems such as an upper foot with a worn center or a chipped edge.

If you need more accuracy, or if visual inspection reveals a lack of parallelism and you want to measure it, use this method: Place a precision wire of 0.010" or 0.020" diameter under the front edge of the upper foot, and zero the indicator. Then retract the contact, place this "master" under the rear edge, and repeat the measurement, noting the variation. Repeat for left and right sides also.

Some gages have machine screws to adjust the reference table into parallel with the upper foot. On those gages that do not, the common solution is to file down either the boss that holds the indicator to the frame, or the shoulder against which the reference table support is clamped. Easy does it!

Make Sure Contacts Are Not Contaminated

This is a common source of error. As in any gaging procedure, the contacts must be kept clean of dirt, lint or hair. When measuring materials such as polyethylene film, which may have been in liquid form only minutes before, beware of the buildup of chemical deposits on the contacts. As with any textile product, frequently clean lint off the contacts.

Check for Gage Repeatability By Checking Several Points on the Surface of a Known, Consistent Sample

Most gages for thin materials do not need to be mastered. Simply "zero" the gage against the reference table. To avoid having to count dial revolutions when measuring thicker materials like acoustic tiles or carpet, use a gage block to set the zero at the nominal dimension of the material.

If a gage reads "under," the problem is most likely a lack of parallelism. If it reads "over," dirt is probably the culprit. Of course, other factors common to all gaging, such as friction in the mechanism or a loose fixture, can also be at fault.

Using And Measuring Precision Balls

Balls are an integral part of every machine shop. They are found in bearings for both rotary and linear motion applications in machine tools; they act as contacts and pivot points in gaging equipment and tooling; and they serve as masters for both size and roundness gaging.

Balls are especially useful to check the parallelism of surfaces on gages with flat anvils and sensitive contacts, including many thickness gages and snap gages. The procedure is simple enough: Pairs of measurements are taken with the ball at the anvil's 3:00 and 9:00 o'clock, and 6:00 and 12:00 o'clock positions, and the anvil is adjusted accordingly.

But in order to perform this function, it is first necessary to confirm the dimensional accuracy of the ball. As with most dimensional standards, the accuracy of a precision ball should be roughly ten times greater than the resolution of the gage which it is used to master, and the instrument used to measure the balls should be ten times as accurate as they are. Consequently, balls are commonly measured to microinches, or even fractions of microinches.

Comparator gages used to measure balls must not only measure in millionths—they must also measure to millionths. They must be designed to the highest standards of stability and precision. As shown in Figure 1, the lower anvil of a ball gage is adjustable for parallelism with the sensitive contact. A frictionless mechanism may be used to ensure constant gaging pressure, and a V-notched saddle may be installed on the anvil to locate balls, with the help of a few degrees of backward tilt. The gage must be in a controlled environment appropriate for microinch measurements, with measures taken to guard against contamination and thermal influences.

Some users like to force the ball between the contacts, reasoning that the pressure tends to "wipe" away any dust or oil that could skew the measurement. Others prefer to retract the upper contact with a lifting lever before placing the ball on the stage, believing that this ensures consistent gaging pressure between trials and reduces stress to the mechanism. The jury is out on this one, so take your pick.

Now, what are we measuring for? When we specify a ball, we are usually concerned with its nominal diameter (D): the value by which it is identified, e.g., $1/2$", 12 mm, etc. Within the range of its manufactured tolerance, this should coincide with the ball's single diameter (D_s), the distance between two parallel planes tangent to the ball's surface—in other words, the results of a single gaging trial.

No ball, however, is perfectly round, and no two balls are perfectly identical. When using balls as gage masters, it is important to establish the level of

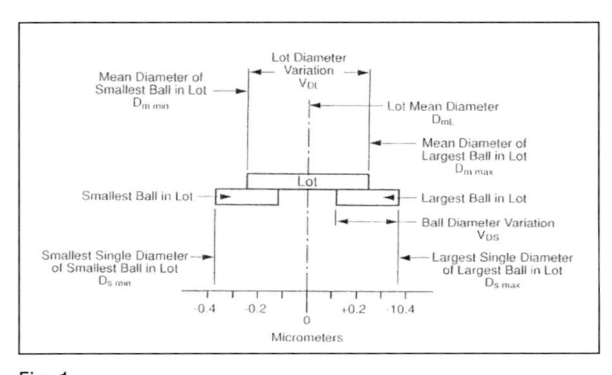

Fig. 1

uncertainty. To accomplish this, individual balls are measured several times, at random locations, to find minimum and maximum diameters. The difference between these, known as ball diameter variation, is simply calculated as follows: $V_{Ds} = D_s \text{ Max} - D_s \text{ Min}$.

Manufacturers of balls, and OEMs who purchase balls as components, utilize several additional parameters for quality control, and to sort balls into quality groups. Another basic parameter is ball mean diameter (D_m). This is the arithmetic mean of the largest and smallest single diameters of a ball, calculated thus: $D_m = (D_s \text{ Max} + D_s \text{ Min})/2$.

To account for production variation between units, manufacturers typically measure balls in lots of ten. Calculations of lot mean diameter (D_{mL}) and lot diameter variation (V_{DL}) are based on the mean diameters of the largest and smallest balls in the lot, as follows:

$D_{mL} = (D_m \text{ Max} + D_m \text{ Min})/2$
$V_{DL} = D_m \text{ Max} - D_m \text{ Min}$

See Figure 2 for the relationship between individual balls and lot measurements.

All of the above are static measurements that can be performed using straightforward comparator gaging equipment. Bearing manufacturers, who require orders-of-magnitude more data in order to predict performance under dynamic conditions, perform additional measurements using circular geometry (roundness) gages. These can generate least-squares circles, analyze waviness geometry in a velocity-proportional fashion, and perform harmonic analysis to predict the "noise" effects of part geometry at various speeds and under various loads. These methods are beyond the scope of what we can cover here. But for most machine shops, who use balls mainly as measuring standards, highly precise comparator gages can provide all the data necessary to ensure accuracy.

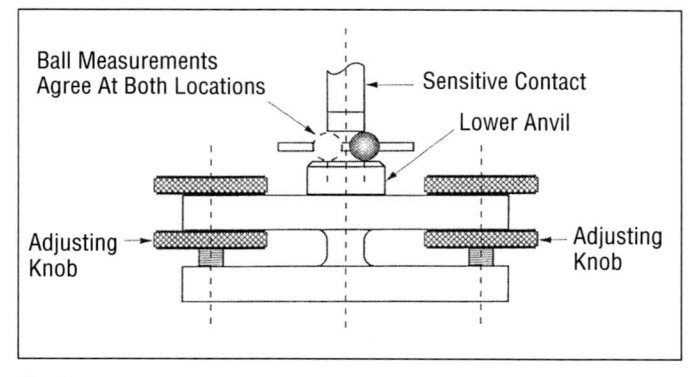

Fig. 2

■ ■ ■

Temperature Compensation

Temperature variation is one of the most significant sources of gaging error. As manufacturing tolerances get tighter and the margin for gaging error gets smaller, it becomes an issue that must be addressed.

Most materials expand as they heat up. For every inch of steel, a 1°F increase causes expansion of approximately 6μ". For brass and copper, the figure is 9μ", and for aluminum, 13μ". If the objective of the inspection process is to determine a part's true size, its temperature must be known. Based on the ISO's very first standard (ISO 1 issued in 1931), that temperature is automatically assumed to be 68°F (20°C).

[Flow diagram: (Workpiece temperature − Workpiece coefficient) / (Master temperature − Master coefficient) / (Gage fixture temperature − Gage fixture coefficient) → Compensation Algorithm → Correction for thermal error]

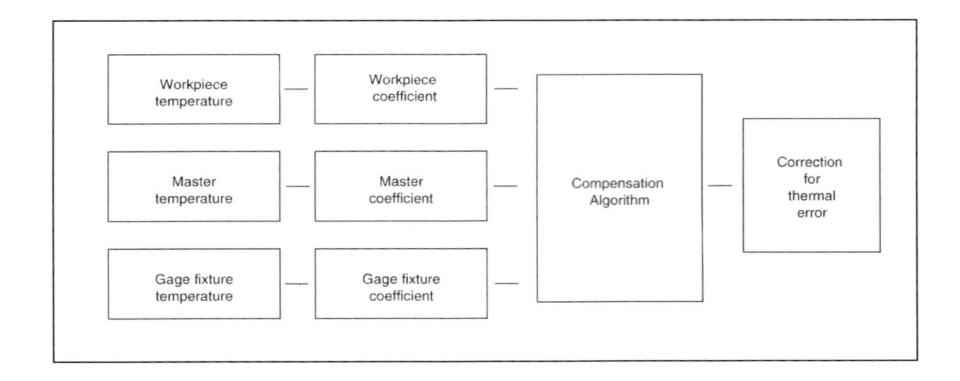

But few inspection processes monitor, much less attempt to control, workpiece temperature. Many quality managers assume that any thermally induced variation in the part will be matched by like variation in the gage and the master: everything will expand and contract at the same rate, and everything will work out just fine.

This is far from true. Gage, master, and part—the three hardware "components" of a gaging system—may be of different materials, so the effects of thermal expansion will differ even if they're all at the same temperature. And the components won't necessarily be the same temperature. Parts that have recently come off a dry machining process may be several degrees warmer, and may remain so for hours. Parts machined under coolant may be cooler. The gage or the master might be sitting on a bench in direct sunlight, or under a heating or cooling vent. Temperature stratification within a room may create temperature differences between components placed near the floor, and components placed on a high shelf. The relative masses of the components may make a difference. For example, an engine block may take longer to reach equilibrium with ambient temperature than a bore gage. And in some instances, thermal variation may work in opposite directions for the gage and the workpiece, compounding rather than canceling the error. For example, high temperatures will cause bore gage contacts to grow longer, which will naturally result in ID measurements that are smaller than actual. On the other hand, the ID of a thin-walled part, like a bearing shell, will grow larger with higher temperatures.

These errors can be significant. As an example, let's use an aluminum part with a critical dimension of 4.0000", and a steel master to zero the gage. The shop is hot today, but both the part and the master are at equilibrium at 80°F (or 12°F above "standard"). Master and workpiece have expanded as shown:

Steel Master: 6μ" × 4 × 12 = .000288"
Aluminum Workpiece: 13μ" × 4 × 12 = .000624"

Error caused solely by the different coefficients of thermal expansion for the different materials is .000624" - .000288" = .000336". That's significant in most operations.

Now assume instead that shop temperature is a perfect 68°F and that master and gage are both at ambient, but the workpiece just came off the machine, and it's 80°F. The entire .000624" variation in the part will show up as measurement error!

Some companies attempt to control this problem by trying to control the environment. Typically, this involves installing sophisticated HVAC controls and building modifications; allowing time for workpieces to reach equilibrium prior to gaging; and other elaborate measures. This works in metrology laboratories, but it's futile in machine shops. The buildings are too large, with too much surface area and internal volume, too many heat-generating devices cycling on and off—too many variables altogether. Essentially, you would have to turn the shop into a controlled lab environment—and that would be really expensive!

A better approach is to measure the temperature of the part, master, and workpiece, and compensate for thermal variation based on the known coefficients of expansion. This is now practical on a production basis using special devices like those from Albion Devices, Inc., (Solana Beach, CA), which interface with electronic gaging systems. Typically, two small, industrial-hardened sensors are installed on the gage: one to measure the temperature of the gage itself, and one to measure the workpiece or master when either is staged. The system can be programmed for different coefficients of expansion of the various components, and the results are fed into an algorithm, which generates a temperature-compensated measurement result on the gage readout. (Additional compensation factors may be built into the algorithm, to correct for unusual elements in gage geometry, differences between a workpiece's surface and interior temperatures, and similar variables.) Such a system will typically reduce thermally induced errors by 90-95 percent. In our second example with the aluminum workpiece, that would bring the error down to 62µ" or less—a figure most shops can live with.

■ ■ ■

Inspecting Tapers
Part 1: Certifying The Master

Conical parts, such as machine tool tapers (i.e., tool holders), gas petcocks, and the shanks of modular prosthetic joints, must often be inspected for taper accuracy. This is usually performed with a special air or electronic gage, custom-made for the specific part. The gage is configured as a female counterpart of the conical part, with air jets or electronic probes located inside at two or more known heights. The gage essentially measures two or more diameters on the tapered feature, calculates the difference between them, and expresses the results either in angular units (e.g., degrees/minutes/seconds or thousandths of degrees), or as deviation from a specified slope (e.g., change of diameter in inches, per foot of length).

Tool holders are of particular interest, because the accuracy of the taper affects the quality of the parts they are used to manufacture. According to ANSI standard B5.10, V-flange tool holders are built with a specified rate of taper of 3 $\frac{1}{2}$" per foot, +0.001"/-0.000. ISO standard 1947 defines a number of taper grades, and establishes different tolerances depending upon both grade and taper length.

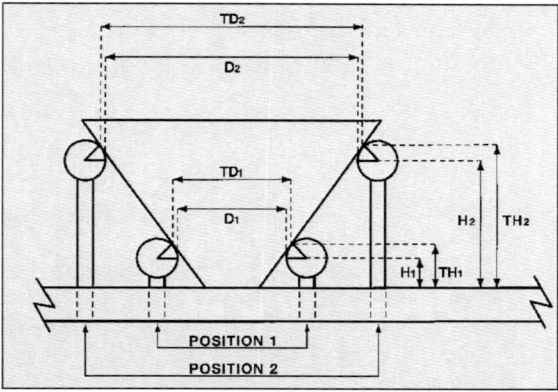

Section N: Applications   227

Regardless of which standard is followed, it is necessary to master the gage before it can be used to measure parts. The taper master is typically a more precise version of the part but, before it can be used to master the gage, it must be certified.

ANSI's 0.001"/ft. tolerance seems easy enough to achieve until you look at the complexity of the inspection process. First, most toolholders are much shorter than 1 foot, so most gages actually compare diameters that are just 3" or 4" apart. Taking 3" as an example, the part has to meet a gaged tolerance of $0.001" \div 4 = 0.00025"$. Using a standard 10:1 ratio, the gage master should be accurate to 25 microinches, and the gage should resolve to the same amount. To certify the master, again at 10:1, will require a ring and disc comparator or a universal measuring machine (UMM) that resolves to 2.5 microinches. A controlled laboratory environment is essential to achieve that level of accuracy.

Certifying the master roughly replicates the production measurement process. The diameter of the master is measured at two known heights, and the slope or angle is calculated from the results. The comparator or UMM is fitted with ball or roller contacts, respectively, whose diameter is known. Because straight-sided gage blocks are used to master the gage (both vertically and horizontally) for measurements on a tapered part, trigonometry is required to calculate the differences between the points at which the mastering occurs, and the points at which the gage contacts touch the tapered part.

Using gage blocks, a very precise indicator, and a height stand, set the height of the tops of the contacts so that the contacts will touch the master at the same height as the lower contacts in the production gage (TH_1 in the diagram). Next, use gage blocks to master the distance between the two contacts, again accounting for the "error" between the mastering points of contact (at D_1), and the measuring points of contact (at TD_1). Then measure the taper master at TD_1.

Using the same procedures, reset contact height and spacing in order to measure the master's diameter at the same height as the upper contacts in the production gage (TH_2). Calculate the angle or slope of the master from the data obtained. Note that the master must be staged absolutely vertical on the comparator or UMM. Prior to certifying the master, a circular geometry gage could be used to check the squareness of the master's end face to its axis.

Certifying taper masters is not for the faint-hearted. Operating within the realm of single-digit microinches, it requires the use of special instrumentation and methods in a lab environment, and a command of trigonometry.

Once the master is certified, however, it can be used to "zero" the production gage, whose use is somewhat easier.

■ ■ ■

Inspecting Tapers
Part 2: Toolholder Gaging

We have discussed the calibration of conical taper masters, which are used to master taper gages. Now let's look at the parts those gages are used to inspect: toolholders.

The most common type of toolholder for CNC machining is the CAT-V or V-flange type.

CAT-V toolholders are external tapers, typically available in five common sizes: 30, 40, 45, 50 and 60. These numbers define both the gage line diameter and length. All sizes have the same included rate of taper of 3 $1/2$" per foot.

There are many reasons for the popularity of V-flange toolholders. One advantage is that they are not self-locking, but are secured in the spindle by the drawbar—an arrangement that makes tool changes simple and fast. They are also economical, because the taper itself is relatively easy to produce, requiring precision machining of only one dimension: the taper angle.

The toolholder must properly position the cutting tool relative to the spindle and, when secured in place, must rigidly maintain that relationship. The accuracy of the tapered surfaces on both the toolholder and the spindle is, therefore, critical.

If the toolholder's rate of taper is too great, there will be excessive clearance between the two surfaces at the small end of the taper. If the rate of taper is too small, there will be excessive clearance at the large end. Either situation can reduce the rigidity of the connection, and cause tool runout, which may show up on the workpiece as geometry and/or surface finish errors. Taper errors may also affect the amount of clearance between the flange on the tooling and the face of the spindle, which may create errors of axial positioning.

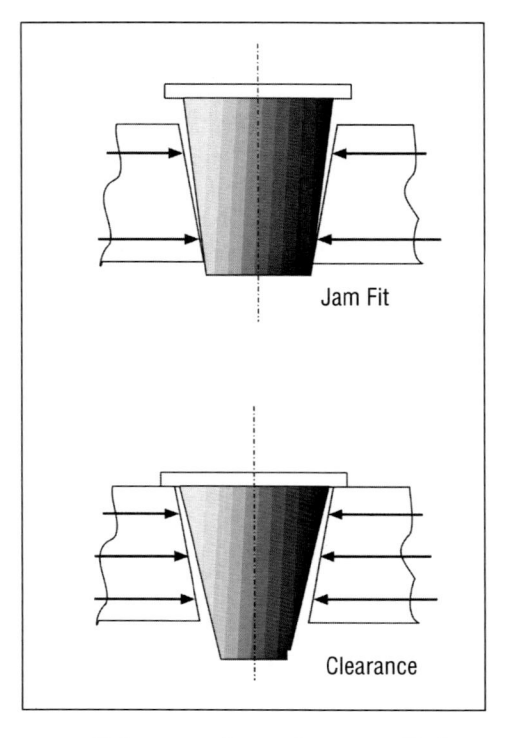

As the demands for precision machining and high speeds increase, manufacturing tolerances on spindle and toolholder tapers have gotten tighter. Nevertheless, both components are still subject to manufacturing inaccuracies and wear. In response, some companies with very high accuracy, quality, and throughput requirements—particularly in the aerospace and medical fields, and some automotive suppliers—regularly check the accuracy of toolholder tapers. This is usually done with differential air gaging, which combines the necessary high resolution and accuracy, with the speed, ease of use, and ruggedness required on the shop floor.

The most common type of air gage taper tooling has two pairs of jets on opposing air circuits, and is designed for a "jam fit" between the part and the tool. Jam-fit tooling does not measure part diameters, *per se*. Rather, it displays the diametrical difference at two points on the workpiece, as compared to the same two points on the master. If the difference at the large end of the taper is greater than the difference at the small end, as shown, the air pressure in circuit A will be lower than in circuit B: the gage will indicate this as negative taper. If the difference at B were greater than the difference at A, the gage would read positive taper. But because a differential air meter displays diametrical difference only, it will not display the part's diameter at either location. So while this type of air tooling provides a good indication of taper wear, and allows us to predict a loss of rigidity in the connection, it does not tell us anything about the tool's axial positioning accuracy.

For that, we need a "clearance" style air tool. The tool cavity is sized to accept the entire toolholder taper, while the toolholder's flange is referenced against the top surface of the tool. This makes it possible to measure diameters at known heights (in addition to the change in clearance, as in the jam-fit type). An additional set of jets may be added, as shown, to inspect for bell-mouth and barrel-shape—two more conditions that reduce the contact area between the toolholder and the spindle.

Given a basic understanding of how your air gage works, both types of tooling are easy to use. Mastering is simply a matter of inserting the taper master and adjusting the zero. Measuring is even easier: just insert the part and take the reading. Be careful when handling the heavy toolholders. Although the air tooling is very rugged, it's not totally impervious to damage.

Gaging Distance Between Hole Centers

Many of the gaging applications we've considered over the years involve size inspection of a single feature, e.g., the diameter of a hole, depth of a groove, height of a gage block, etc. Many parts, however, contain multiple features that establish dimensional relationships between two or more other parts. Examples include engine blocks, with multiple cylinder bores that establish distances between pistons, and pump housings with two overlapping bores that establish clearance between two mating impellers. As with people, machined features tend to be simple and straightforward when they're single. When people or parts are in relationships, however—and especially when those relationships involve mating—things can get complicated.

Gaging the distance between hole centers is a good example, because it requires that you first locate those centers in space before you can measure the distance between them. Gaging equipment may vary widely with the application, and include considerations of part size and configuration, and required throughput and accuracy. But the basic idea remains the same across various technical approaches; you must first establish references between the part and the gage, often in three dimensions, before you can measure deviation from a specified distance between features.

A basic approach is a hand-held gage, like that shown in the figure. Both of the plugs have two fixed contacts, and one spring-loaded contact each, so they automatically center themselves in their respective holes. One of the plugs is a fixed reference; the other is a sensitive contact. Note also the hard depth stops, establishing a reference in the third dimension. The sensitive plug moves relative to the fixed plug on a frictionless device such as a pantograph mechanism or an air-bearing carriage for good repeatability. The sensing/indicating device used to measure deviation may be a simple dial or digital indicator, or an electronic probe and amplifier.

This gage is economical, reliable, and easy to use. It may be employed with parts as large as an engine block (or larger), or as small as a connecting rod, and its capacity may be adjustable to measure different distances. The plugs may be replaceable, to accommodate different size holes. Its limitation is that it only measures distances between centers; no other types of features or relationships may be checked.

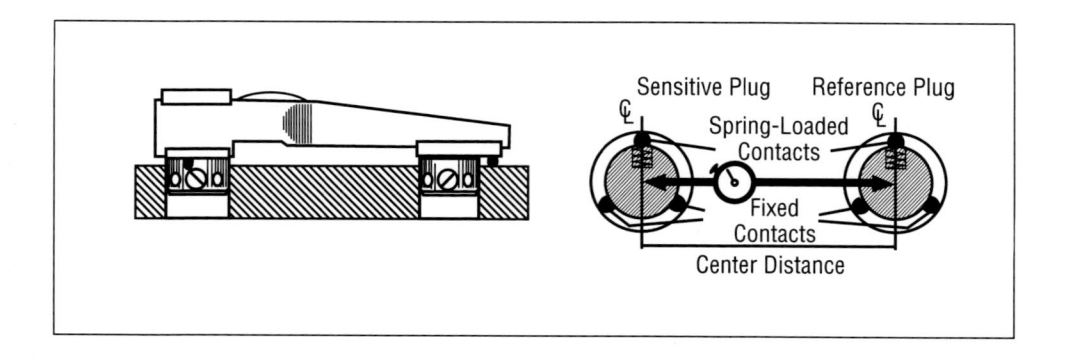

The next option is a gaging fixture engineered for a specific part. Such gages are generally benchtop devices, so they are limited to use with relatively small parts (e.g., automotive conrods) that can be brought to them. The workpiece is typically located over fixed plugs, each of which may contain one or more air or electronic probes as sensing devices. Fixture gages may offer limited adjustability, but tend to be very application-specific.

The benefit side of the coin, however, is substantial. Fixture gages tend to be very stable, and allow high throughput rates. Because the probes can be spaced quite densely within the fixture, it is possible to measure multiple features or characteristics simultaneously. For example, a conrod gage with sixteen probes (eight each for the wrist pin and the crankshaft bores) can be engineered to measure the following features, in addition to center distance between bores: four diameters per bore (6:00 to 12:00 o'clock, and 3:00 to 9:00 o'clock, at both top and bottom), bore "out of roundness" (i.e., the difference between two diameters at right angles), bore taper, and bend and twist between bores. Gaging computers and some amplifiers can be readily programmed to perform all these measurements without changing the gaging setup.

For large, mass-produced parts with multiple features in relationship (such as engine blocks or cylinder heads), an elaborate "doghouse" gage may offer the best combination of accuracy and throughput. The gage gets its name from its large, cast-iron structure that contains machined reference surfaces and "sweep" gage plugs, and establishes precise relationships between them. The engine block is pushed into the doghouse, and reference surfaces on the workpiece are positioned against matching references in the gage. The sweep gages are single-probe gaging plugs that are mounted in bearings, and may be rotated 360° on their axes to establish the locations of the hole centers. Operation may be either manual or automatic.

In an automated gage, multiple sweep gages descend simultaneously into the appropriate bores. A gaging computer then takes the data and calculates the multiple relationships between the bores. This may even include not only distances between adjacent bores, but also distances between opposite banks of cylinders on V-configuration engines. Results may be presented in numeric and/or pictorial displays.

Obviously, such a gage is costly, and must be engineered for the specific application. It is often justified, however, in automotive and similarly high-volume/high cost applications where multiple relationships must be measured to high levels of accuracy at very high rates of throughput. And in essence, it's not very different from the simple hand-held gage; the basic idea is still to provide a means of reliably locating the workpiece against fixed reference contacts, then measuring deviation relative to those references.

■ ■ ■

Gaging "Relational" Dimensions

Specifications often require inspection of dimensional relationships between two features, or between two dimensions on the same feature. For example, a TIR (total indicator reading or total indicated runout) specification defines a relationship between the outside diameter of a part feature, and the part's axis. The taper of a conical part is another relational specification, which is checked by comparing diameters at two known heights. Other examples include distance between centers, ID or OD ovality, and parallelism.

Compared to single-dimension specifications (such as diameter), these "relational" specifications can be challenging to inspect. Not only do they involve multiple measurements; they also require that separate measurement results be somehow combined in a mathematical relationship. This might be as simple as subtracting the minimum reading from the maximum reading, as in the case of ovality, or so complex as to require trigonometry, as in the case of taper measurements.

Of the three standard gaging methods available for these tasks—indicator gaging, electronic gaging, and air gaging—air possesses the best combination of properties for most applications. Dial indicators are somewhat bulky, restricting the close spacing of contacts. And to combine the results of two or more measurements, the operator must usually perform mathematical calculations. While electronic gaging amplifiers can be readily programmed to combine signals from multiple contacts and calculate results directly, many electronic gage contacts are too long to be placed in tooling for inside dimension measurements.

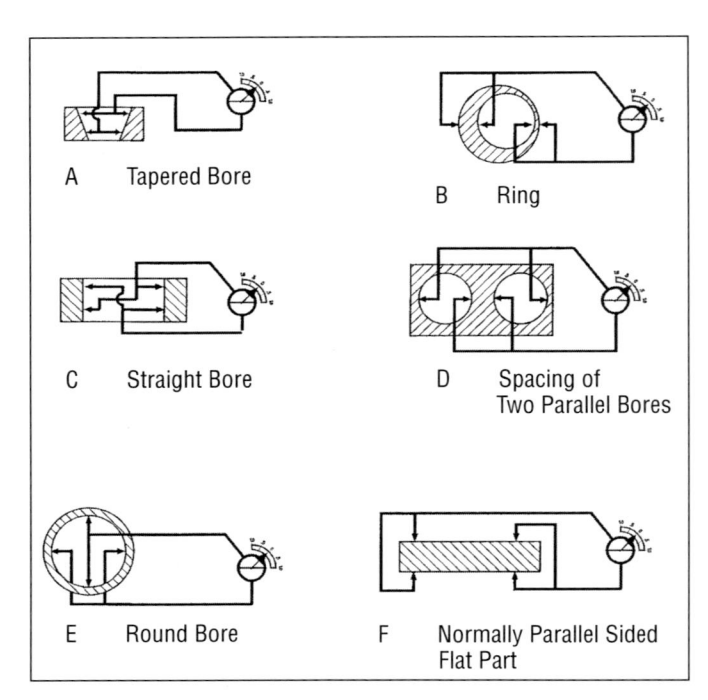

A   Tapered Bore
B   Ring
C   Straight Bore
D   Spacing of Two Parallel Bores
E   Round Bore
F   Normally Parallel Sided Flat Part

The type of air gage known as a dual-circuit air comparator (also known as a differential, or balanced, gage) has a circuit working on either side of a bellows in such a way that the gage reads zero when pressure in both circuits is equal, while any differential between the circuits is displayed as variation from zero. With appropriate tooling, this arrangement allows the gage to combine signals from multiple jets, to generate relational dimension results directly. In addition, air jets are quite small, permitting great flexibility in gage design, especially for inside measurements.

The diagrams show schematically how jets are arranged in tooling to measure various relational dimensions. As shown in (A), a conical taper gage places pairs of jets on separate circuits, to compare two diameters at known heights. Gage (B) measures wall thickness variation, or eccentricity of ID to OD. The tooling in (C) measures bore squareness, while ignoring bore diameter and taper. Two such tools could be combined in a single fixture to measure bend and twist of two bores relative to one another, or bore parallelism in two planes.

Gage (D) measures the center distance between bores by comparing distances between opposite walls of the bores, while ignoring changes in bore diameter. Gage (E) measures ovality by comparing two diameters in the same plane, at 90 degrees to one another. And gage (F) measures parallelism by comparing part thickness at two locations.

Another common "relational" inspection task (not shown) is the measurement of clearance between mating parts, as in fuel injector and bearing assemblies. We can't squeeze a gage between the assembled components to measure clearance directly, but we can measure two unassembled parts simultaneously, using one gage circuit for the ID component, and the other for the OD component. The gage will thus display the "match" or clearance. Should it indicate that clearance is out of tolerance, we can leave the OD part on the tooling, and inspect a number of ID parts until we find a proper match.

It is often useful to obtain simple, single-dimension measurements simultaneously with relational measurements. For example, in addition to inspecting wall thickness variation in (B), we may also want to measure the outside diameter. Air gaging excels at this task. In all the examples, additional jets could be installed in the tooling, adjacent to the ones shown. These new jets would run on separate air circuits to a second comparator, to perform simultaneous measurements of diameters (in diagrams A through E), or thickness (F). In many cases, the tolerance for the "relational" dimension will be significantly tighter than the tolerance for the "simple" dimension. If this is the case, we can specify different levels of magnification for the two gages, each one suited to the range and resolution required.

Beyond The Height Gage And Surface Plate
What Can Take Low-Volume, Precision Inspection To The Next Level?

For many years now, the method of choice for low-volume, general-purpose inspection has been surface plate work using test indicators and height gages. Recently, electronic height gages have made the layout inspection process a little more accurate. Speed of measurement has also improved, since electronic height gages give direct measurements and allow for storing data and programming repetitive measurement processes.

Are there even more advanced electronic height gages on the horizon that will allow us to achieve greater precision while still maintaining our ability to do a wide variety of measurement tasks? Probably not. The problem is not the height gage, but the reference surface it rests on.

No matter how precisely we build the height gage, the accuracy of the measurement is still dependent on the flatness of the surface plate. Surface plates probably cannot measure up to the increasing demands for higher tolerance measurements.

So where do we go from here? Consider the Universal Measurement Machine or Universal Length Gage as a way to perform a wide variety of measurements and inspections with speed and improved accuracy. While there may be no all-purpose machine that can do everything, a universal length gage can do quite a lot. Applications include:

- Internal and external measurements of diameters and lengths (master rings and discs)
- Internal and external thread measurements
- Calibration inspection of mechanical indicators and gages (dial and digital indicators, LVDTs, test masters)
- Location of points, lines, holes and surfaces
- Internal and external tapers

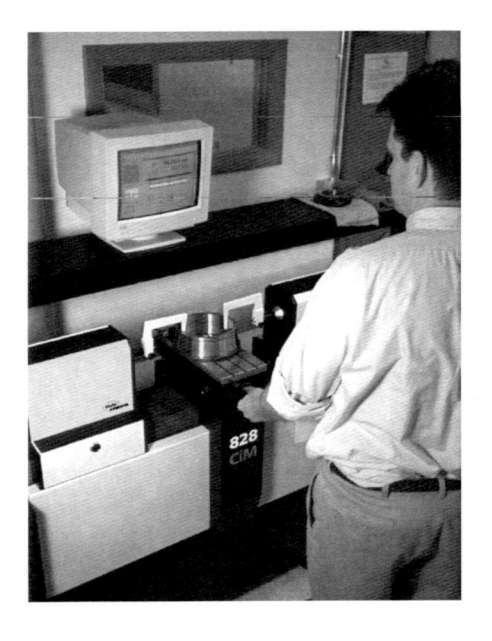

Universal measurement machines were developed to speed up the inspection process and reduce the potential for measurement error. They differ from typical comparative style gaging because they have a much larger measuring range, but still can obtain resolution and accuracy approaching some comparators. In order to achieve high measurement performance, the machines have built-in reference standards—either glass scales or an interferometer system.

When equipped with various contact accessories, universal measurement machines can easily be used to check length, diameter, pitch diameter, roundness, straightness, parallelism and taper. They will typically measure parts from 5 to 40 inches long, but machines are also available with even larger capacities.

These rugged looking systems are frequently referred to as "machines" because they are built according to the same design criteria as machine tools. Critically important measuring head and tail stock slide bearings are mounted on a strong and rigid base. The reference system is mounted as close as possible to the machine's line of measurement to avoid abbe errors. In addition, various computer techniques are used to map and correct slide errors; average multiple, lightning-fast measurements; and compensate for temperature variation.

No, it's not a surface plate and height gage; but a machine that is extremely fast, versatile and very accurate. However, just like anything else, measurements made with the machine are only as good as the measurement process. Therefore, it is important to keep all the components of the process the same when setting up (i.e., measuring the exact same location on the part, verifying gaging pressure, standardizing on a contact style, ensuring the utmost cleanliness of the part and contact, etc.). Machines do this with extensive computer-aided systems that help you set up a measurement process and then lead subsequent users through it the same way every time and, of course, capture and report measurement results and analyses.

The height gage and surface plate have been a mainstay for highly productive, low-part-volume general purpose measurements. The universal measuring machine incorporates the spirit of this process, while relying on a highly accurate internal reference to significantly improve measurement precision and repeatability. It is really an extension of the same line of thinking that has served manufacturers so well for over 100 years.

Deep Thinking About Depth Gages And Their Evolution

A depth gage is a very common hand tool used to inspect the depth of holes, slots, counterbores, recesses or the distance from one surface to another. They are especially common in the tool and die industry. Like other hand tools, they have undergone a gradual change from mechanical scales to digital wonders.

In the beginning a depth gage was simply a scale with a sliding head. The scale was set into the hole and the slide squared up with the reference surface. The depth of the hole was then read directly from the scale itself. This was a simple tool, but it did not absolve the operator of his responsibility to employ good judgment and proper technique.

To ensure consistent measurements with any kind of depth gage, it is important to adhere to some basic ground rules: Make sure the cross bar or reference head is clean, flat and free of nicks and burrs. Hold the manual gage flat and square to the reference surface. Any out of squareness of the head to the surface introduces error. So if you aren't careful, you may be measuring along the hypotenuse instead of the actual depth of the hole.

As tolerances increased, the depth scale gave way to the vernier depth gage. This type of gage took a little longer to read, but with some training and experience the users benefited from greatly improved resolution.

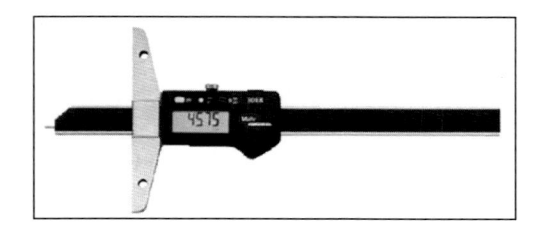

Both the scale and vernier type depth gages need to be set to zero. This is done by placing the measuring head on a flat surface, such as a surface plate, and moving the sliding arm or contact to the same surface. If the reading on the tool is not zero, it should be adjusted so that it is. Unfortunately, the sliding members of both the scale and vernier depth gage are quite large and are not suitable for measuring small holes (e.g., $1/4$" or less).

Another improvement was the micrometer depth gage. It uses the barrel of the micrometer as the measuring leg and allows for entry into smaller holes. Micrometers provide for absolute rather than comparative gaging. Micrometer depth gages achieve very good resolution over their entire range; however, the range itself is limited. Therefore, for deep holes, an extension and a master are required. When using extension rods, always remember to add the length of the extension rod to the measurement value shown on the micrometer.

The digital caliper may also be used for making occasional depth measurements. Most incorporate a depth extension as a standard feature. The depth extension may be square or round, and the style of choice is determined by the hole size you need to measure. For the smallest of holes (less than 2 mm), round is the way to go. The only problem with using a caliper as a depth gage is that it's not what the gage is really designed to do. It is very difficult to align and hold the caliper square while making the depth check. The end of the caliper is just not designed for stability.

Of course, the fastest and most accurate tool for checking depth is the digital depth gage. Made in various sizes, digital depth gages may be fitted with a number reference for covering larger diameter spans. There are even extensions for spans up to 300 mm/12". Digital calipers are also self-supporting on the part and have sufficient width for good stability.

There are a number of different configurations for the actual depth contact. One version of these gages uses a small, fixed contact on the end of the slide to measure the depth. This is fine for general-purpose applications where the opening is large. Another style uses a rod with a replaceable contact. This offers a lot more versatility, allowing the user to change contacts according to the surface that needs measuring.

The moral of this story: There are many ways to check a depth dimension on a part. However, if speed and accuracy are important, use a tool that was designed for those purposes—today it's the digital depth gage.

3, 2, 1, Contact Measuring Thickness

Thickness is one of the most frequently measured dimensions, and also one that is very easy to understand. So you might think that someone would come up with a one-style-fits-all measurement approach good for just about every kind of thickness application. But it just isn't so. There are many approaches to measuring thickness, depending on the requirements of the part.

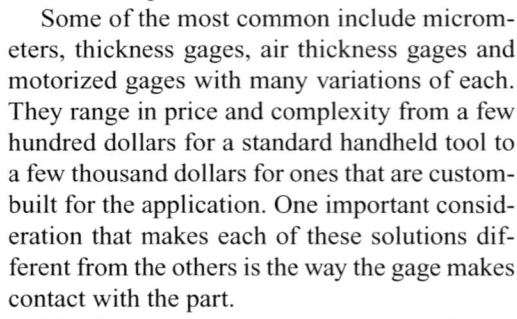

Some of the most common include micrometers, thickness gages, air thickness gages and motorized gages with many variations of each. They range in price and complexity from a few hundred dollars for a standard handheld tool to a few thousand dollars for ones that are custom-built for the application. One important consideration that makes each of these solutions different from the others is the way the gage makes contact with the part.

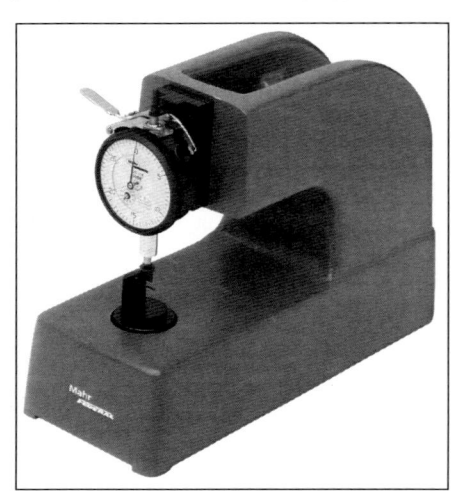

Whether it's the thickness of a piece of sheet metal, a silicon wafer, a latex glove or photographic film, we are usually talking about measuring the distance between two parallel surfaces. Accurately determining the length of a line that is perpendicular to two parallel surfaces has everything to do with how the gage makes contact with the part.

Micrometer. A handheld micrometer is a simple low-cost method for measuring the thickness of a piece of sheet metal, for example, which is relatively stiff and thick. It certainly provides a lot of measurement range. With flat and parallel contacts and constant gaging force applied with the friction or ratchet drive, the micrometer can self-align to the part for a fast and accurate reading. A potential problem with the self-aligning flat contacts of the micrometer is that they can bridge across and "average out" minute variations in thickness. So if higher resolution is required, you should look for a different approach.

Thickness Gage. The portable thickness gage raises the ante on resolution by combining a flat anvil-type reference surface on the bottom with a radiused (ball) measurement contact on top. Making a single point measurement eliminates the possibility of gage error that could be caused by faulty parallelism of the contacts. This would be a good gage for a narrow strip of photographic film where a spot check of thickness is required.

Air Thickness Gage. In some cases the part may be so susceptible to scratching or marring that any amount of gaging contact could destroy the part. An air thickness gaging system directs thin, precision-aligned, opposing streams to each side of the part, which is held perpendicular to the air streams on a ground base. The gage measures backpressure on each of the air streams. Since the backpressure is directly proportional to the distance between the contact point and the nozzle, it is easy for the gage to automatically calculate thickness. This non-contact technique provides a means of sliding the part around for measuring thickness variation. The air provides enough cushion to help float the part as it is repositioned. The use of differential probing provides fast, accurate measurement regardless of where the part is positioned. Most importantly, the part is not damaged by the gage contacts.

Custom Thickness Gage. Many soft, compressible parts have rigid, standardized specifications for how they are to be measured. In such cases the design of custom gages will frequently take into consideration the size and shape of the contact along with the gaging pressure applied to the part. Usually the gaging pressure is defined as a dead load weight to assure constant force over the full range of measurement. However, with some compressible materials, such as paper or latex rubber, the amount of time that the load is applied to the part will also affect the reading from the gage. In these cases, the gage may incorporate a motorized measuring contact along with the cam actuated device to retract the contact after a specific period of time. This uniform load and dwell measurement prevents the parts from deforming and setting.

As you can see, the method of probe contact is a very helpful way to consider how to set up thickness gaging systems across a wide range of applications. Choosing the right approach can dramatically decrease the chance of inadvertently making a bad measurement.

The "Issues" With Height Gages

Don't tell any one, but there is something of a problem with height gages. The big issue that they have is what they measure, height. The larger the height gage, the bigger the potential problem.

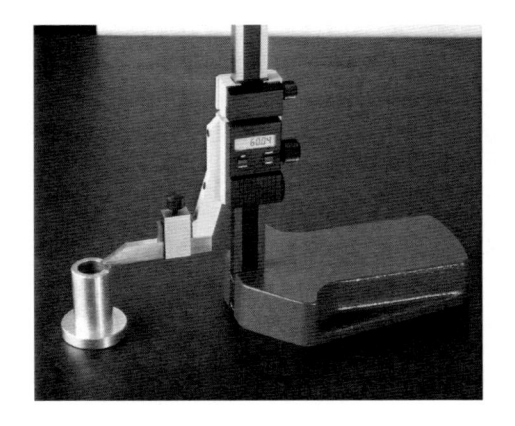

It's not actually the height that is the problem. It's the relationship of a large height to the small base. Just like a lever—the longer the arm, the larger the multiplying factor—and with a height gage this is the problem. We are not only talking about the errors coming from the gage itself, but also errors in the setup. These get magnified and can potentially distort an otherwise carefully planned comparison.

A major error in the design of a basic height gage is taking a design that was meant to measure 12" and simply extending the post to measure 36", without changing the base design or the cross-area of the measuring post. What then naturally happens is that the gage will tend to wobble and flex. Although you may not be able to see the 0.001" wobble, it can become a significant part of the part tolerance and certainly influence the measurement.

A normal step in trying to increase the performance of the gage is to beef up the column in an attempt to reduce the flexure of the post. However, this only gets us part of the way to a better gage. For example, if a slight pressure is placed horizontally against the gage's measuring contact, the gage may slide along the table. If the same force is applied to the measuring contact when it is near its maximum upper position, this force will very likely cause the gage to tip over. What needs to be done is make the base longer and wider, and build in some mass. By decreasing the ratio of the post to the base there will be significant improvement in its performance.

Besides its own personal issues, the height gage tends to keep company with tools having bad references. Most height gages are used with a surface plate. The surface plate provides the reference for the part and the height gage. Many surface plates are clean and well maintained. But there are those that may not be as clean as they look. A small metal chip or even a hair, while almost impossible to see, could throw off the measurement by 0.020" at a height of only 10".

Next to dirt, the actual surface of the granite surface plate will play a key role in the performance of the gage. Any slight imperfection between where the part and the gage are staged will get amplified the higher the measurement. Most surface plates have a flatness spec of 50μ". If the base is say 6" long, a 50μ" error would grow to more than 0.0003" in 36", and even worse if the plate is out of spec.

As a concerned parent to your height gage, it's up to you to deal with its issues and choose who will be working with it. Look at the height gage to make sure it is beefed up for the job and check to make sure it has a precision surface to rest on.

■ ■ ■

An Inside Look At Special Diameters

Sometimes we are faced with making critical inside diameter checks on parts that do not present themselves in a straightforward fashion. Usually these checks are on the inside of some type of bearing and they can be almost any size. Examples include measuring the diameter of an internal surface behind a shoulder, where the entry diameter is smaller than the diameter being measured; or measuring a ring groove in a bore, the pitch diameter of an internal thread, the effective diameter of a barrel roller bearing, or the included angle of a tapered bore.

Measuring this type of ID requires that the gage have some unique characteristics. These include the high repeatability of a comparative gage, long-range jaw retraction to allow entry into the part, and the ability to set the depth of the contact to make the measurement at the proper point.

Many people immediately think of a shallow bore gage for this type of application. As we have discussed in the past, this type of gage rests on the face of the part and measures the inside diameter at a certain depth. These are fairly accurate comparative gages that allow the depth of the contact to be set to a specific location. However, these gages are typically limited by the amount of travel found in the indicating device used for the readout. As a simple diameter gage they work well, but they do not meet the need of long range to get behind a shoulder, or to have the contact measure a groove diameter (Fig. 1).

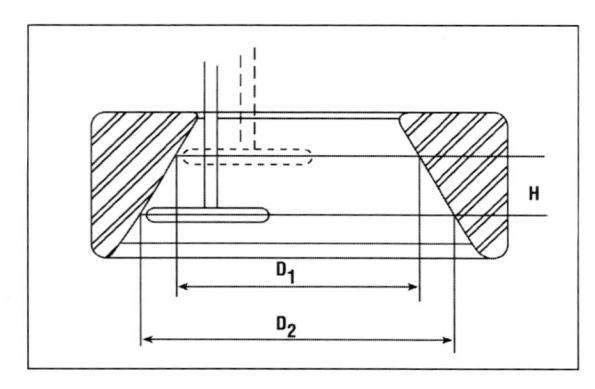

Fig. 1

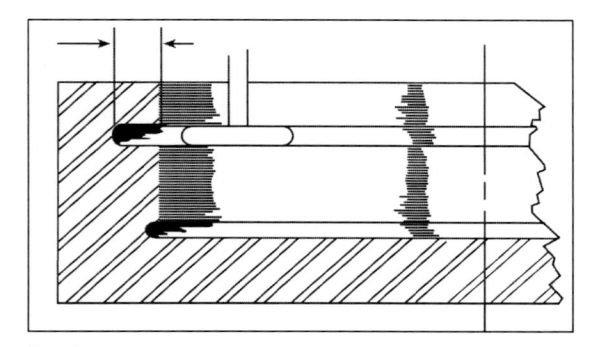

Fig. 2

Two types of gages do meet the needs of these applications: a gage very similar to a shallow, comparative bore gage, but with a long-range slide; and a long-range slide coupled with a high-resolution, long-range digital indicator.

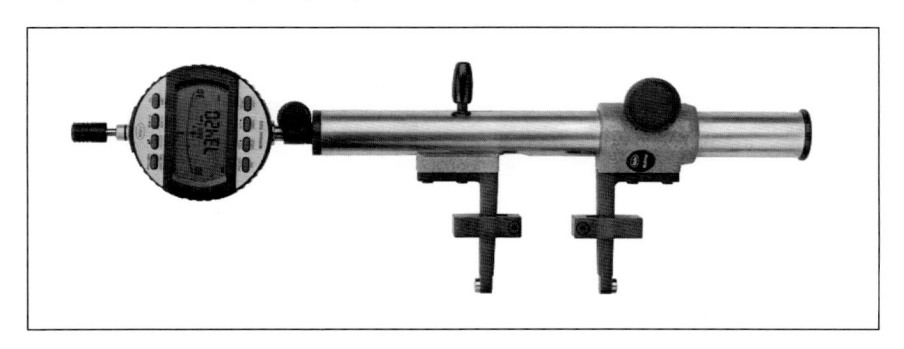

The first example uses the precision slide purely as a transfer mechanism, or a way to retract the comparative indicator away from the surface to get past the obstruction. When released, it slides back against its stop and the comparative indicator does its thing. This sounds like a simple solution until one considers the mechanics involved. The contact has to be long enough to get down and around the smaller diameter. This is where the design of the slide and contact become critical. If the slide has any looseness in it, it will cause the gage to be unrepeatable. Or, if the contacts are too flimsy, they will bend and twist and become sensitive to placement and gaging pressure. Either of these design flaws in the precision slide can make even the best comparative indicator look bad.

A better way of achieving the same result is to use a long-range digital indicator as part of the measuring frame. This has the benefit of actually referencing the measurement based on the accuracy built into the gage's own long-range slide. Of course, the contacts must be designed to handle the depth and gaging pressures, but the beauty of this type of gage is that its long range offers the ability to measure any number of depths within its range. So, besides measuring a depth at a particular location, it can measure two diameters at different depths, thus becoming a taper gage measuring the included angle of the tapered bore (Fig. 2).

The combination of the long-range, high-resolution digital indicator on a soundly designed mechanical frame can provide a universal ID (or OD) gage that has endless applications for those hard-to-reach inside measurement applications.

Holes Big Enough To Fall Into

Maybe you're not in Texas, but suddenly you find yourself faced with a huge measurement requirement. You've been given the task of checking some large diameters… Not your 6" variety, I mean those large enough to drive a herd of cows through. You know, the 12", 36" or even the 80" variety.

Don't go for the tequila yet. There are lots of choices available to meet this challenge, which boils down to selecting the right tool for the application. The first step is to look at the part print, determine the measurement tolerances, and see if there are any callouts for out-of-round conditions. Those two pieces of information will lead you to the best tool for the job.

If the tolerance is loose—within 0.01 inch—then a digital or vernier caliper-style gage will provide a good fast check of the part diameter. Just make sure the jaws are square to the part and placed to find the major diameter. On the larger diameters, this could even be a two-person operation.

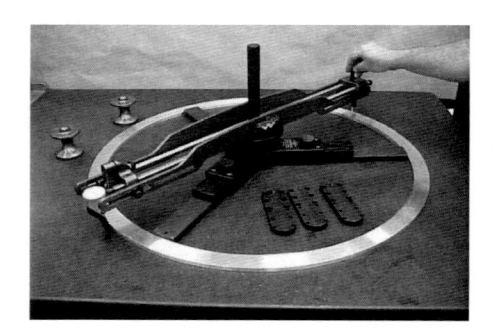

An inside micrometer is another alternative. Special kits make it possible to assemble a series of calibrated extension rods to span any diameter. Because this is a true point-to-point measuring system, the diameter has to be found by rocking the gage both axially and radially. On a large bore, this may require one operator holding the reference side of the gage in place while the second operator "searches" for the maximum diameter.

Tighter tolerances call for different types of gages. Some adjustable bore gages can get to these larger sizes. They deliver improved accuracy and repeatability because they: 1) are adjusted to a specific (in this case large) size range; 2) provide comparative measurements using a master; and 3) are often equipped with a centralizer which makes it easy to "search" for the diameter. Pair this gage with a good digital indicator, which includes a dynamic function to store the maximum size, and you have a great tool for fast, repetitive readings.

Gages with beams that have reference and sensing contacts mounted on either end are another comparative tool. In addition to satisfying large diameter measurement requirements for tolerances within 0.001 inch, beam type gages have standard rest pad and contact combinations that allow measurement of shallow bores and thin wall parts, as well as grooves and other features machined within the bore. You can even build up the gage to get around a central hub.

When the blueprint requires you to check not only the diameter but out-of-roundness as well, the bar has been raised. The gages mentioned above can still be used, but the process may involve making five or ten measurements on the part, recording the results, and calculating out-of-roundness according to a formula. Not only is this approach time consuming, it also magnifies operator influences on the result because so many measurements are required. An advanced concept can be brought into play here: the better the gage is staged, the better the result of the measurement. That's why the shallow bore gage with its two references is better than the gages that just have one reference. By the same thought process, another reference point will result in even further improvement. Let's take the same shallow bore gage, but this time we will use it with a staging post that centralizes both the part and the gage. The operator only has to apply a little force to make sure the reference contact is against the part, and the central post takes care of finding the maximum diameter without having to rock the part back and forth. Now it's a breeze to inspect for out-of-round conditions.

Just rotate the gage, keeping a little force applied to the reference contact and watch the swing of the needle, looking for the Min and Max values. Watching the needle, you can visually inspect for the Total Indicator Reading (TIR) or out-of-roundness variation. Add a memory to the indicator, or an amplifier to store discreet points, and you can automatically calculate the average roundness.

By building on solid measurement concepts, staging the gage and the part, and eliminating operator influences, you can easily reduce your Texas-style large diameter measurement problems down to the size of Rhode Island.

Improving Height Gage Results

As with any measurement, the quality of the result depends on the measurement instrument and the care with which the operator handles the measurement procedure. Many gages are designed to make this as easy as possible. A snap gage, for example, has the reference anvil, frame and measuring instrument built in. The same is true with a bench stand. To obtain a good measurement, all the operator really has to do is correctly apply the gage to the part.

With an electronic height gage, this is not quite the case. Electronic height gages are not as self-contained as other gages. They have the precision scale and the sensing probe, but no integral reference, which is the most critical part. Most electronic height gages are used on a granite surface plate and the plate provides the reference for both the height gage and the part that is being measured. The quality of this base plate directly influences the measuring result. Thus, it is important to keep the plate free from dust, chips and dirt.

Most height gages are direct reading instruments that generally operate in ranges up to 36 inches. As such, they are especially susceptible to variations in temperature. Since the body heat of the operator (98.6°F) is clearly above the room temperature (68°F), any heat conveyed to elements of the measuring circuit (base plate, test piece, height measuring instrument, stylus) can cause local heat expansion and induce measurement errors. Operators should be very careful in observing the following rules:

- Avoid touching the test piece with your bare hands directly before the measurement. Use gloves.
- Do not touch other elements of the measuring circuit.
- Only touch the height gage at points provided for this purpose: handles are usually provided to move the gage or engage the air bearings for positioning.
- Avoid drafts.
- Avoid direct sunlight on the instrument, test piece, or base plate.
- Do not set up the measuring station in proximity to radiators or in the path of air ducts.
- Do not check test pieces that were transported through very hot/cold rooms shortly before measurement.
- For high precision measurements, put the test piece on the base plate and let it adjust to ambient temperature (approx. ¼ to 8 hours, depending on the size of the part).

Once the measuring loop is verified, there are two other critical references that need to be established. The first is the zero-reference for the measuring system. With automated height gages, this is done automatically whenever the gage is turned on. In a manually driven gage, the gage must be zeroed on the granite plate before it can be used. With a motor driven unit, the gage will automatically move down to touch the surface to set its reference point. It's not a bad practice to initiate this zeroing routine a second time, just to make sure that no dirt or other anomaly has introduced an incorrect reference. Since setting this reference is critical to all the measurements you will make, it is certainly worth the time and effort.

The other important reference is the correction for probe ball diameter. If a height gage is to be used only for length measurements taken with the probe moving down, then probe diameter is not important. The contact point of the probe will be the same as in zeroing. But, if grooves, diameters, or the hole locations are being measured, or if any measurements are taken with the probe moving upwards, then the probe ball diameter must be known and taken into account.

Ball diameter is specified for the probe, of course, but there is always some degree of variation. Actual ball diameter should be added to any dimension that is probed in the upward direction.

On height gages that have even the most basic electronic control this dimension can be measured as part of a set-up routine and is automatically included in all measurements. The automated process uses a fixture provided with the gage, or the test can be simulated with a couple of gage blocks. The fixture sets up a plane that is measured by the gage from both directions. The gage then looks at the difference between the two measurements and calculates this as the ball diameter.

The same gage block check can be done by hand on purely manual machines or the ball diameter may be measured off-line with a micrometer. Just as with setting the zero reference, this check should be repeated a number of times. A lot of gages will provide this repeat check automatically and reject the ball diameter reference if it does not repeat to within a preset limit.

Failing to recheck for ball diameter can be a deadly pitfall when a probe tip is changed. Going from a 10- to a 5-mm ball tip would be disastrous if not recalculated.

■ ■ ■

Air Gaging Styrofoam

Over the years, we have discussed a great many successful uses of air gages in this column. We have seen it used to measure diameters, tapers, straightness, and even little bitty holes. Air gaging has always proven to be an excellent choice for fast, easy-to-use and reliable, high performance dimensional gaging. But did you ever think it could be used to check those Styrofoam end caps that secure your new stereo system in its shipping container? Well, it can, and it does it very well.

Styrofoam, or expanded polystyrene (EPS), is a basic term for styrene polymers that can be expanded into a multitude of different products. EPS has been available for more than 50 years and proven in numerous packaging applications. It is ideal for packaging and shipping applications because of its strength, light weight, cushioning characteristics, dimensional stability, and the fact that it is thermal and moisture-resistant.

EPS is supplied to molders in the form of polystyrene beads that are loaded with a blowing agent, usually pentane, and other agents that give the beads the ability to expand and be molded into low-density foam products. In the end, EPS is 90 percent air—which accounts for its light weight. By varying the mixture of chemicals and the heating process, the foam characteristics can be changed to vary the density and create the best mix for the specific packaging requirement.

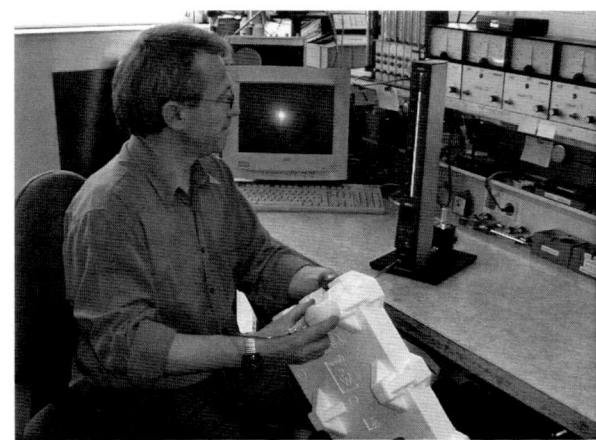

So what does this have to do with air gaging? Because the density can be varied to change the characteristics of the material, the end product then has to be checked to make sure it will perform under load. You have probably noticed differences in some of the Styrofoam products that came with your DVD player or new computer: some are rock hard, while others may be a little "crumbly" and break apart easily. What is at work here is the bonding agent in the foam. Regardless of the desired density, if the bonding agent has been used successfully, the beads will all be held very tightly together. If not, they will tend to break apart—not only in your hand, but during the transport of your precious new toy. Bonding removes air between the beads and seals the matrix.

This is where the air gaging comes in. The less air between the beads, the stronger the bond. The more air, the weaker the bond. And if there is air between the beads, there is also a path for air to flow through. By applying pressurized air to the Styrofoam, we can monitor the flow of air through the material. It's like using air to measure little bitty holes, but in this case we are measuring a lot of itty-bitty holes. Just as we used the air gage as a flow meter for small holes, we can do the same for this material.

In the application, an open-air probe is pushed into the Styrofoam part. A stop collar on the end of the probe limits the probe travel into the part and also acts to seal off any air escaping from the hole. If the bonding of the beads is good, there will be virtually no place for the air to go, which presents high backpressure to the air gage. If the bonding is not so good, there will be gaps between the beads allowing the air to flow through, resulting in less back pressure.

Just as with previous flow gage examples, we need a set of reference standards to set up the low end of acceptability. The gage is then set up with these standards to create the limits. Now it's pretty easy for the operator to use: he brings a part to the gage, sticks in the air probe and observes the scale. Limit lights can also be used to set the acceptable limits.

Another characteristic of EPS is how well it flows through the mold and matches up to the molding surface. If the flow is correct, it will follow exactly the surface of the mold. If the mold has a flat surface, the part will be flat. Should the flow not be correct, the beads will not completely form the surface and the end result would be a slightly bumpy surface that could easily break off and get into the shipment. Again, air gaging can be used to check this by manufacturing an open-air jet probe with a large flat surface, which is laid on the surface of the part. If the surface is smooth, the probe will seal nicely with the part. If the process was not quite right, gaps will exist, allowing air to escape. Both cases represent changes in back pressure that the gage can measure and display.

What is next for air gaging? Well, let's see…how about tapioca pudding?